S0-EKW-732

Canoes the world over

illustrated by **C. Earl Bradbury**

THE UNIVERSITY OF ILLINOIS PRESS, URBANA, 1952

by **Terence T. Quirke**

Canoes
the world over

Ⓤℙ Copyright 1952 **University of Illinois**
Manufactured in the United States of America

Introduction	6
About the Author	7
Chapters 1 The Canoe	9
2 Simple Dugout Canoes	14
3 Poles, Paddles, and Sails	21
4 African Dugouts	35
5 American Dugout Canoes	42
6 Reed Canoes	54
7 Skin-Covered Canoes — Kayaks	65
8 Primitive Skin-Covered Crafts	73
9 Canoes Made of Bark	81
10 Birchbark Canoes	94
11 Sailing Canoes	108
12 Double Outriggers and Double Canoes	117
Glossary	126
Bibliography	128
Index	131

Introduction

My interest in canoes and canoe building started in boyhood. My father was a musician, and my brothers and I were brought up with very naïve ideas of the use of tools. I am sure that many a savage had greater technical skill to bring to his canoe building than we had. However, we did have good cutting tools, and we were able to purchase from local lumber dealers the requisite wood for ribs, gunwales, and end posts. By the time I was seventeen years old I had shared with my older brothers in the making of three canvas-covered canoes, largely of our own design.

From our schoolboy trips it is a far cry indeed to the thousands of miles of exploration on which I paddled in later years. Many a shining lake, many a weary portage, many a rushing river have been traversed, thanks to the convenience, portability, and safety of simple open canoes. But every sportsman I ever met who knows canoes agrees that each of us is only an amateur when compared with primitive man. All over the world primitive men have performed marvels of ingenuity, strength, and endurance with canoes, and have shown a heroism, skill, and mastery of their craft that puts civilized sportsmen to shame.

Some years ago I started to read about the canoes of other lands. I consulted museum bulletins, anthropological and ethnological journals, books of travel, geographical magazines, museum exhibits, journals of missionaries, reports of fish commissions, explorers, and other scientists. In this book I have tried to tell what I found interesting about canoes of many kinds, how they are made, what they are for, and about the men who use them.

Terence T. Quirke

About the Author

This book by my husband, Terence T. Quirke, was written over a period of years, as a result of the persuasion of our friends and our children who delighted in his stories of canoes and canoemen.

My husband's interest in canoes began in his boyhood days when he explored by canoe the tidal rivers of the south of England and continued during his career as a geologist when, for nearly twenty seasons with the Geological Survey of Canada, he paddled canoes over thousands of miles in the north woods of Ontario. A skillful canoeman himself, he delighted in the dexterity and courage of primitive peoples; and with a scientist's eye he gleaned, from previously unexplored sources, interesting facts about primitive watercraft and the peoples who used them.

He considered certain sections of this book — such as the classification of handholds of paddles, and the study of the extinct Yahgan canoe made of the bark of the evergreen beech — as original research contributions.

Professor C. Earl Bradbury, the illustrator, was a friend of long standing. I recall how delighted my husband was when he agreed to do the illustrations. His work with figures has been outstanding, and he was sensitive to the need for accuracy in making the illustrations of the canoes authentic. For photographic reference material, museum models, and drawings by ethnologists, Professor Bradbury is indebted mainly to the *National Geographic* magazine, the Smithsonian Institution, and the United States Fish and Wildlife Service bulletins. He took great pains to make the illustrations technically correct as well as interesting and beautiful.

Although the manuscript was completed before my husband's death, it was set aside for several years; and this delay necessitated help in preparing the manuscript for publication. This help our son, Terence Jr., gave gladly, and I am grateful to him and to our many friends who, like him, have expressed hopes that this book be published.

To Professor Bradbury, I owe more than I can express. His illustrations do all that my husband had hoped, in making vivid the skill and courage of primitive peoples who have built and used canoes the world over.

Anne McIlraith Quirke

Fig. 1. American Indian birchbark canoe.

The Canoe 1

The ancient Egyptians were probably the earliest mariners, and they were followed by the Greeks and Romans. No travels ever had such a telling as those of the Greeks. But all the voyages of Ulysses and his crew, picnicking around the island-studded Mediterranean in ships, were mere week-end parties compared with the journey of twenty-three-hundred miles between Tahiti and Hawaii which was made in Polynesian canoes. If another Homer were to sing the tale of Polynesia, what an Odyssey that would make.

Columbus brought the word "canoe" to Europe. It is a corruption of the native Arawak word learned by the Spaniards in the West Indies, and reported by them as *canoa*. "Canoe" it has been for English-speaking peoples for about four hundred years.

A canoe is a simple craft, made to be propelled by paddles, poles, or sails, narrow in proportion to its length, and having at least some of its buoyancy due to a hollowed shape. Canoes are made of a variety of materials, depending upon the geographical location of the makers. In countries lacking suitable timber, canoes are made of reeds, especially papyrus, and certain rushes. In the Arctic, where there are not even rushes, the natives make their canoes of skins. Some men make canoes with leather or rawhide coverings. Where large trees grow, men make canoes by hollowing out a log, or by stripping off bark. They make a vessel out of nothing but bark, or out of bark stiffened with an inner framework. Other men starting with a dugout canoe build up the sides with boards, palm midribs, or any other material available to them. Others weave a basketwork and daub it over with pitch.

The greatest achievement of all is to set up a mast and sail and be able to travel in such a frail-looking craft for thousands of miles at sea. Canoes sail with a single outrigger, double outriggers, or no outrigger at all. Two canoes may be attached side by side for sailing, or a wide-beamed, roomy dugout can be made staunch and steady enough to carry a sail.

All aboriginal canoemen use the canoe as a means of livelihood. Those who are nomadic, such as the Yahgans of Tierra del Fuego, the North American Indian of the old days, and the Eskimo, needed canoes in order to permit their families to follow the distribution of their food supply. Many use their canoes for getting fish. Others use canoes in hunting, in transporting native products, and in carrying on warfare or escaping from its consequences. For almost all aborigines the making of canoes is a serious and laborious business, not to be undertaken for caprice or amusement, and no native tribes ordinarily have any spare canoes. The men usually patch and cobble their old crafts until they will no longer hang together.

Human settlement in many parts of the world has depended on canoes. In North America, where Indian canoes penetrated ever farther and farther north and west as the Indians were driven back by demands of the fur trade and by European encroachments, the birchbark canoe has great human interest. (See Fig. 1.) Similarly, the natural spread of population along the waterways, from the coast lines into the green wilderness of tropical America, makes the Amazonian canoes of great importance. When the enormous area of the Pacific Ocean and the spread of outrigger canoemen to almost every island from Alaska to New Zealand and from Madagascar to Easter Island are considered, what the canoe has meant to mankind can be better appreciated.

The white man has contributed to canoe building because of the tools and materials at his command, but all his adaptations of leeboards, outriggers, balancing boards, and sails have been anticipated, at least in principle, by some aboriginal canoeman. Even the models he prefers today, the so-called "flatboat," and the common open canoe are copied from the Eskimo and the North American Indian. The white man's sailing canoes are surpassed in speed by the Johore fast boat. His leeboards were anticipated by the ancient South American raft. His balancing outboard was copied from the balancing platform used by the Polynesian sailors of the Gilbert Islands. Even his watertight compartments might have been inspired by the sealskin floats of the Chilean paddler. He clings slavishly to the single-pole mast, whereas the native canoeman has used single, double, and tripod masts.

Even in craft used for recreation, the Hawaiian islander gave the white man a splendid lead in the invention of the surfboard. No canoe races can equal in excitement those of the Maoris who actually slide over hurdles in canoes. For splendor and holiday zest the Siamese and Chinese dragon-boat races cap the climax of all hilarious regattas. Given equal opportunities with tools and materials, the native canoeman the world over can beat the white man hands down. Every fair-minded observer has praised the superb skill and courage with which aboriginal canoemen handle their craft. To courage and skill in operation they have added amazing versatility in design and construction.

Which is the best canoe made? That question is impossible to answer adequately. There are so many different types, made of so many different materials and used for such diverse purposes, that there must be many *best* canoes and many *best* canoemen. Certainly some tribes make better canoes than others out of similar or identical materials, although some tribes have better natural materials. The tribes that had access to birch bark made much better canoes than those who had to use elm bark. In Guiana the Djuka, Bush Negroes, make much better dugout canoes than do the aboriginal Indians, although they use the same materials. In the basin of the Congo, with similar materials available to a large number of tribes, some natives make excellent dugout canoes, some produce such poor canoes that they are not even straight, and others use nothing but rafts. The tribes that make the best canoes, in general, have the most skillful canoemen.

What is the best way to paddle? There are certainly many poor ways of handling a paddle. Paddles may be too short, too long, too narrow, or too wide for effective use by some individuals, but there is no best size or design. If there were any best paddle theoretically, every skillful aboriginal canoeman might be expected to use the same sort of paddle. Actually there are considerable variations in paddles, but they are variations within limits. Obviously, however, the double-ended paddle is better for the Eskimo kayak man than a single-bladed paddle. There is similarly a great variety in the use of paddles — standing, sitting, or kneeling — but every way is a good way when well done.

Canoes express the man. The course of human culture can be traced in the evolution of float to raft, raft to dugout, dugout

Fig. 2. Ancient Egyptians constructing papyrus-bundle canoe.

to built-up canoe, and thence to modern ships. The earliest known dugouts were used by the prehistoric lake dwellers of Europe, the oldest skin-covered craft were coracles, the oldest ship contours followed the pattern of the ancient Egyptian papyrus floats (see Fig. 2), and archaic sails were used on the sailing rafts of Brazil before the days of the Incas. Some of these primitive craft were not really canoes, but their bearing upon the way in which men have developed canoes has brought them into our story.

There could not be much purpose in a book about canoes unless it linked them to man. Even the Great Spirit, according to Indian lore, added the birchbark canoe as an afterthought to his favorite creature, man.

Fig. 3. Raft of mangrove logs, western Australia.

2 Simple Dugout Canoes

A Canadian lumberjack standing on a log floating down the river can make a log look easy to ride. Whenever an experienced lumberjack goes downstream on a single log without the support of other logs beside him, he does so either because he can't help it, or out of bravado. Those who think that the simplest craft possible is a plain log, which a rider may paddle along with his hands and feet, know little about the subject.

In the course of exploration I have often had to cross rivers without a boat. A single log is so apt to tip that it is the last thing I would choose as a float to ferry across my clothes, watch, compass, and notebook. I usually tried to find three small dead trees, one of which had an upstanding branch or root. When these were lashed together, they supported clothes and instruments above water on the upstanding snag, and yet were steady enough so that they did not turn over as I swam behind and pushed them across the river. The superior stability of this contraption depends upon the width of the three poles lashed together, as compared with the width of a single log.

People all over the world use rafts of some kind, varying from very crude floats to elaborate barges supporting a house. The Australians furnish the classic example of men who can ride a log and paddle with their hands. Drawings show Australian men sitting on a log, with legs outstretched and feet supported on the log while the rider paddles with both hands. To those who have not seen the feat, such a report is incredible.

The Australians are also said to ride on rafts made of several logs bound together and on bundles of rushes. Figure 3 is drawn from the photograph of a native of the Buccaneer Archipelago, whose home is an island broken off from the northwestern coast of Australia. This man has solved the problem of stability by making a raft of mangrove logs. It is true that the craft will not turn over, but it is also true that its value is negligible. The craft is very primitive, but so is the man. In spite of his magnificent physique — he stands nearly six and a half feet tall — he has a very limited mental scope. He can, however, capture great sea turtles and harpoon sea cows.

A raft offers the stability that a single log lacks. Let the log be flattened on one or both sides, and its stability is greatly increased. Although a plank is not likely to capsize, it will not float a rider without becoming partly submerged. In other words, a plank or flattened log lacks adequate buoyancy. Chopping or

Fig. 4. Shilluks in dugout, Khartoum, Sudan.

burning out the middle part of the log will lighten the resulting shell to about fifteen per cent of the original weight and increase the buoyancy of the log. The hollowed log rides much higher in the water and is in fact a crude canoe.

A dugout may be defined as a log lightened by removal of the center. The simplest dugouts are made from palm trees, which have a soft pithy interior. In India they make a craft of this sort called the "dunga," which is common in the rivers of Bengal. The Shilluk tribe on the Nile occasionally use a crude dugout made from a local palm. Figure 4 shows several men in such a dugout. It is as round as the original trunk and square at both ends, the pithy interior not being suitable for shaping to a point. It is slow, apt to tip, and generally unsatisfactory, but it does have more buoyancy than a solid log and fair stability when a heavy load is placed close to the bottom. In this canoe even the Shilluks, who are wonderful canoemen, find it advisable to sit on the bottom and stay in the middle without rocking the boat.

The labor of hollowing out a log is hard and time-consuming for primitive people. Imagine cutting down a tree with a stone axe and hollowing it out with a stone adz, pumice stone, and fire as tools! Nevertheless, canoes are hewn out of logs nearly everywhere that navigable waters are found, and various methods have been devised to make the process as easy as possible. Some canoe-makers drill or burn holes about an inch or so deep from the outside, and use these holes as guides for hollowing out the inside. With the help of these holes they can cut down the sides as thin as desirable, without the danger of cutting them too thin. This device has been invented independently by peoples as far apart as the Indians of Guiana and the Samoyeds of Arctic Siberia.

After the interior of the log has been chopped out and after the round bottom has been cut away sufficiently, the sides may be left so low that they fail to keep out the splashing waves. The problem now is to combine stability and buoyancy with adequate freeboard. The freeboard of any craft is the height of the lowest part of the gunwale above water level.

Many peoples solve this problem by (1) choosing very large logs, in order to secure high sides for the canoe; or (2) by spreading the sides of the hollowed log into a wider, shallower shape — in most cases spreading can be done only after the wood has been softened by soaking in water.

In order to facilitate spreading, the Arawak and Carib Indians of South America cut out both ends of the log, making it merely a trough. Later they plug up the open ends with V-shaped end pieces, or they force the ends to curve up above water level. The Cayapa Indians of Ecuador cut down trees of great size and hew from the logs the outside shape of the canoes as planned. They follow a design which provides high bow and stern, with a flat and much lower central part. Finally, they burn and chop out the inside to conform to the lines already hewn on the outside. The high ends produce a definite upward curve of the gunwale at each end of the canoe, called the "sheer" of the craft. Its purpose is to prevent swamping by splash over bow and stern; it serves to add extra freeboard where it is usually badly needed.

The Haida Indians of the North Pacific Coast used very large pine trees or the giant cypress for the hulls of their canoes, cutting away as little as possible from the tops of the gunwales, and then building on a high cedar bulwark at bow and stern to prevent swamping over the ends. Peoples outside of North America have added sideboards, wash strakes, above the gunwales all around the canoe to keep out wave splash. However, when a dugout is built up with structural additions, it belongs among the more elaborate types of canoes. By the logical development of superstructure above the dugout as a foundation, many modern boats have been evolved.

In order to attain speed and ease of handling, the canoe must have lines different from those of the original log. The primitive canoe must be fined at each end by cutting the square ends to a point or to some other modified shape. The cross section of the canoe must be reduced gradually from a wide, flat part amidship to a narrower, streamlined bow and stern. Canoe-makers smooth and polish the craft on the outside, in many cases using a varnish, grease, beeswax, or native paint to reduce the friction against the water. A smooth surface and fine lines add not only beauty but speed to the finished canoe.

The fastest canoes of the world are the so-called war-canoes. These are characterized by great length, small beam, and a large crew of paddlers. The form of the war-canoe illustrates one of the limitations which always face the canoe-maker. He can gain length for his craft by using a long log, but he cannot increase the beam much beyond the diameter of the original tree. On the other hand, the European boatbuilder increases the beam of his boat in proportion as he increases the length, maintaining a ratio of not less than one to ten between the two. The canoe-maker, however, in various parts of the world produces a craft over fifty feet long and less than three or four feet wide. For the canoeman, great length has certain advantages. In a long canoe stability per man is increased, provided all the men do not move in the same direction at the same time. Propulsion power is greatly increased by the additional paddlers, and water resistance is relatively reduced. These advantages make a racing canoe out of the long dugout, and

Fig. 5. Maori war-canoe, New Zealand.

also make it one of the fastest man-driven boats in the whole world.

War-canoes go so fast that their paddlers easily outdistance the highly trained rowing crews of European warships. There is record of a British officer in Burma, who, using such a canoe with thirty paddlers, traveled a distance of forty-eight miles in six hours. Great speed has always been required for the war-canoes of savage tribes, who have made them deep, narrow, and long, with a sharp sheer fore and aft. These swift canoes enable the native generals to move the greatest number of men to any place on water at the greatest speed. For paddling speed there is no other primitive craft which can be compared with the long narrow dugout. In addition, the high stem and stern protected the crew from being raked by flying arrows or spears in encounters with enemies.

Dugouts have been used all over the world, in ancient and modern times. Thousands of years ago Julius Caesar brought his troops to Britain in ships. Paleolithic man in the Stone Age had made dugouts when his only cutting tools were flints. Chinese records of 2000 B.C. express the idea that all ships were derived from the dugout type. The ancient Greeks, Romans, and Egyptians thought that the papyrus ark was the original ship. It is possible that there are Eskimos who believe all ships started as skin-covered kayaks, and that there are Indians on the Amazon who think that all boats started as woodskin or bark canoes. No matter what the materials available, man has always had to face the problems of buoyancy, stability, and propulsion of his watercraft. How these problems have been met and to what extent they have been solved by the canoe involves a world-wide study.

Fig. 6. African canoemen standing to paddle.

Poles, Paddles, and Sails 3

Poles are the most obvious and perhaps the most primitive method of driving canoes along. In many localities they are effective, convenient, and easily obtainable. For hunting or fishing, it is found that poles are more nearly silent than paddles and more effective in turning the canoe about and in bringing it promptly to a standstill. In fishing at sea the Papuans go out at times in a small one-man dugout. These dugouts are so narrow that when a man sits amidship, his thighs overlap each gunwale and all but touch the surface of the water; such canoes would be useless without an outrigger to steady them. The Papuan paddles noiselessly along until he comes to the edge of the coral reef. Without a sound he changes paddle for pole; and standing with one foot on each gunwale, he punts along slowly and silently. Suddenly he brings his canoe to a full stop, puts down his pole, raises his pronged spear, and steadies himself for the throw at an approaching fish. The spear flies into the water, the canoeman stoops for his pole, recovers his spear with a fish wriggling between its prongs, throws the fish into the dugout, and prepares for another cast.

In central Africa many of the Ubena tribe earn their living

by carrying rice in large canoes using poles. Similarly, on Lake Chad the natives pole along their canoe-shaped rafts made out of bundles of papyrus reeds. They use canoes as freighters for carrying domestic animals, family possessions, and other native baggage about the lake, and they carry especially, as a regular trade, quantities of salt. Poling is possible anywhere on Lake Chad; although over 10,000 square miles in area, it is nowhere more than fifteen feet deep.

When a canoeman can plant one end of his pole on solid bottom and press steadily on the other end, he can make good progress even with a heavy craft. In many native craft the canoemen walk back toward the stern as they push on their poles in the manner of a European bargee. Natives who navigate weedy and muddy streams or swamps are likely to use poles. The Seminole Indians of Florida push their canoes through the weedy water or through mangrove-lined watercourses, using poles exclusively.

The Creole frog hunters of Louisiana, who stalk their quarry through very shallow places in the cypress swamps, push their tiny, shallow dugouts with the end of the same pole they use to catch the frogs. Even in Europe, which we seldom think of as the home of native canoes, we find the country folk of the Pripet marshes of Poland using a flat-bottomed hand-hewn canoe which can be poled along in only a few inches of water. In fast and rocky rivers, poles are much safer and better to use than paddles. Furthermore, a pole saves the danger of splitting a paddle blade against a rocky river bed. Needless to say, canoemen who have poles also use paddles as they are required.

The dugouts on the upper Ubangi River, the most northerly of the great tributaries of the Congo, are punt-shaped, with flat, square ends and flat bottoms, suitable only for shallow waters. The canoes are divided into sections, generally three or four, by partitions four to six inches above the bottom. These partitions are made when the hull is first fashioned from a tree trunk. The bow is reserved for the steersmen who stand up to use long polelike oars or sweeps. Behind the steersmen, the main part of the crew use poles from nine to sixteen feet long. They press on the bottom of the river and push the boat by walking along the platform within their own section. In addition to polers there are paddlers who sit in the stern and use little paddles only thirty inches long to help the steering of the navigators in front. These canoemen, who call themselves "Watet" or water-folk, have worked out a system by which the short-handled paddle, the long-handled paddle, and the pole are each applied to best advantage at the same time in the same canoe.

Most canoes, of course, are propelled with paddles. A double-bladed paddle usually makes for greater speed than a single-bladed paddle. It is, however, much more tiring and in narrow streams very awkward. For use on open water, either a large lake or the sea, a double-bladed paddle is far better for one man in a canoe. The canoeman, seated near the center of the canoe, can with a single stroke of the paddle turn his craft into the wind or into position to take approaching waves at best advantage. He has good control of his craft, even when exposed to violent or sudden gusts of wind from which he can have no shelter or relief on open stretches of water.

A single Eskimo canoeman in a skin-covered kayak can easily outspeed two men in an ordinary open birchbark canoe, and he can also easily get away from men skillful in the use of rowboats. This great speed and agility of handling is due not only to the use of a double-bladed paddle, but also to the very light and handy form of the kayak. On the other hand, the Samoyeds of western Arctic Siberia, used the double-bladed paddle in the one-man kayaks but single-bladed paddles in the two-man kayaks, bidarkas. All the Aleutian Eskimos, however, used double-bladed paddles in their two- and three-man bidarkas. Perhaps in imitation of the Aleutians, the Amur River canoemen used double-bladed paddles in birchbark canoes holding from two to four men.

The double-bladed paddle is used by few native races except the Eskimos. Very few of the Polynesians, fewer of the Africans, and none of the Asiatics, except those near or in the Arctic, use a double-bladed paddle. Certainly many tribes and races have occasion to venture far out into the open sea, but most of them do not operate as lone hunters as much as the Eskimos do.

In South America the only regular use of double-bladed paddles was that of the natives who used sealskin craft along the Peruvian coast. Chile is over ninety degrees of latitude away from Alaska and on the other side of the Equator, but the canoemen of both regions independently developed a sealskin craft. Although their canoes are radically different in design, both tribes ventured out on the open sea and used double-bladed paddles.

The single-bladed paddle is generally handier for all canoemen on rivers, lakes, and sheltered sea coasts. Convenience should control the usage of paddles and canoes. However, this is not always the case. Primitive peoples and highly civilized peoples, too, heed the heavy hand of custom and tradition with very little regard to logic. Because a certain tribe uses a specific type of canoe or a special design of paddle, it does not necessarily follow that it is the best conceivable for the purpose. In many cases the type of canoe and paddle used have been thrust upon the tribe by force of tradition or superstition.

Among primitive peoples, ceremonial is closely related to tribal law and government and must be accepted as a factor of very great importance. In general, a ceremonial object is one which has a mystical part in some public ritualistic observance. Such objects have no utilitarian purpose, but often are fashioned into ornamental or grotesque copies of common and useful objects.

Ceremonial paddles are made by many of the tribes of the Greater Pacific. In Africa, Nigerian ceremonial dances are sometimes directed with a highly ornamented paddle, intended for no other function. Also among some Congo tribes, the chief or headman holds a carved and otherwise ornamented paddle as the insignia of his office. In North America, the Indians of the northwestern coast seem to have used certain paddles for ceremonial. The ancient races in South America, long before the coming of Europeans, had elaborately carved paddles which are presumed to have had ceremonial uses.

Ceremonial paddles tell nothing about the methods of using canoes, but they do indicate the degree to which canoes and canoeing have entered into the basic culture and have moulded the outlook of certain peoples. Wherever ceremonial paddles are found, it is pretty safe to say that the tribe has an ancient and intimate connection with canoe culture. It follows that the race has long been dependent upon watercraft, probably for its very existence.

There are very different ways of paddling. These depend upon the position of the canoeman, who may stand, sit, or kneel. These various postures control not only the length of the paddle handle, but also determine the way in which the paddle is held. The North American canoeman nearly always sits or kneels to paddle, which makes a relatively short paddle handle desirable. The paddle is held with the lower hand grasping the handle around the shaft, palm of the hand down, thumb around opposite to the clasp of the fingers, and with the upper hand placed over the end of the shaft in such a manner that the palm presses against the handle during the stroke. This has led to the general adoption of an enlarged handle at the end of the canoe paddle, and in some North American tribes, especially those on or near the western coast, to the development of a definite crosspiece at the end of the handle. A paddle with a crosspiece at the end is also very generally used in the Dutch East Indies, Malaya, China, and in various parts of Polynesia.

In Ecuador there is a curious distinction between the women's paddles and the men's. The women use an enlarged-end handle, but the men have ornamental ends on their paddle, frequently spear-shaped, which makes an end hold for the paddle practically impossible. I suspect the underlying reason for this difference is utilitarian. The men usually stand to paddle, and the women habitually sit on the bottom of the canoe while paddling.

In Africa we find occasional use of another paddle hold, the "both palms down" variety. The Kimwani tribe, living on the southwest coast of Victoria Nyanga, use very good dugouts and plank-built canoes. The Watet of the upper Ubangi River operate in fast-flowing shallow waters; but the Kimwani carry on their fishing with canoes on a great open lake, where they are in danger from swamping and from being overturned by waves. All the

Kimwani canoemen sit down to paddle. The steersman at times uses a long sweep for steering, but ordinarily uses a paddle like those of the crew. He sits on a raised stern piece, overlooking the paddlers who sit on racks of wood slung from the sides of the canoe.

The paddles are made from a single piece of wood, each about four and one-half feet long with a spoonlike, circular blade about fourteen inches across. In paddling, each man takes great care to use the concave side front; that is, the pressure is against the convex side of the paddle. This is very important to the paddler, because otherwise the paddle twists in his grasp. The Kimwani grasps the shaft near the butt with one hand, and with the other he holds it about two-thirds of the way down. The shaft passes across both hands, and both hands are palm down. Each hand grips with all the fingers around the shaft and the thumb in the opposite position. During the stroke the shaft of the paddle is held in a nearly vertical plane, just as with the North American canoeman, even though the hand hold is essentially different.

The steersman, whose seat at a slightly higher level gives him a good view ahead and over the paddlers, acts as coxswain and sets the time of paddling by a rhythmic chant. This set tempo he respects himself, even when upon occasion he changes his paddle from side to side without losing stroke. The singing of the paddlers is typically African.

All over equatorial Africa, canoemen sing; a leader chants the verse, and the crew joins in the chorus. The words are usually monotonous repetitions, cheering the workers with remembrance and anticipation of pleasures. Occasionally, however, the songs are crudely comical, and African laughter swells the chanty chorus. Natives who have learned gospel hymns at the missions embellish the English words with African variations, and results are ludicrous beyond description and often weirdly sacrilegious.

The both-palms-down hold is, in general, adapted either to those who use a double-bladed paddle or to those who sit near or below the level of the water. The Tarascan Indians of Mexico, for instance, use this hold when they are in their small canoes, but they use the North American hold when they are in their large canoes. This is because they fish in very small dugouts and sit on the bottom. The very small canoe sinks considerably in the water under the weight of the fisherman; consequently, he has to use his paddle at a level considerably above that at which he sits.

The one tribe or group of tribes that uses a straight-ended paddle handle that I know about in North America uses the Kootenai type of bark canoe in British Columbia. These people seem to have used a grip around the handle of the paddle with the upper hand opposite in position from that of the lower hand. This is the palms-in-opposition hold. It is a grip widely used in Central and South America, throughout Polynesia, and in Africa.

The Africans, in general, use a long-handled paddle and stand to paddle their large canoes. (See Fig. 6.) There is no doubt that a barefooted African can get a much better footing in a dugout than can a European shod with any sort of shoe. While paddling at full strength, the African paddler seems almost to grip the bottom of the craft with his feet. This enables him to put the full strength of his whole body, with a greatly increased reach, into his work. To do so he must use a rather long paddle. His paddle almost always tapers slightly at the upper end, and he grasps the paddle with the palms-in-opposition hold. When standing to paddle, the forward reach comes from the swing of the body as far forward as the balance upon the feet permits, the backward swing of the body ends when the weight of the body reaches the position vertical above the rear foot; from that point on, the paddle stroke is accomplished by a bending forward of the body as the arms continue the final kick of the long, very powerful stroke. The wider the reach of the arms, the greater is the power of that final push.

Some of the African paddlers, especially on the Congo and its waters, increase the reach of their arm spread by about two inches, using a curious upper-hand hold. In this grip the paddle handle passes between the index and the longest finger of the upper hand. This results in pressure being applied at the far end of the palm of the hand instead of at the near end, as is usual with the hand hold over the end, or with palms-in-opposition around the paddle handle. This grip is peculiar to Africa, and probably only to certain tribes. It is a grip useful only to paddlers who stand, and only to those who must be prepared for sudden bursts of power beyond that used in routine travel. The African paddle grip, as an emergency speeding hold, probably came about due to the danger of sudden attacks from rushing crocodiles, rogue hippopotamuses, or slave hunters.

The Papuans of the interior of New Guinea are among the best exponents of the long-handled paddle. These people use long and very narrow dugouts on the rivers and inland lakes. Theoretically, they ought to sit as close to the bottom of the craft as possible in order to prevent their upsetting. On the contrary, men and women stand to paddle. Like most paddlers who stand, they grasp their paddles with hands-in-opposition position. They use paddles with handles eight to nine feet long, in canoes perhaps thirty feet long and less than twenty inches wide.

When twelve or more men are paddling, the middle portion of the long canoe can actually be lifted from the water by the rhythm of the paddlers. As the paddler puts his weight into the beginning part of his stroke by bearing down on his paddle, he relieves his own weight on the bottom of the canoe. Since the ends of the canoe are solid and much heavier than the center, they tend to sink while the center rises. At the middle of the stroke, when the power is applied backwards, the paddlers brace themselves securely by their footing in the canoe, and all their weight comes down again; at that stage the descending arch of the bent canoe violently slaps the water at the end of the stroke. The same thing may be seen as the bow of a Canadian canoe rises and sinks slightly when the paddlers begin or end their stroke, but nowhere else have I ever heard of such a stunt as that performed by the Papuan paddlers.

There is also wide variety in the shape of the paddle blades. Some are circular like those of the Tarascan Indians, the Somali of the Red Sea, the Papuans in the interior of New Guinea, and certain tribes on the Amazon River. The North American paddle blade is usually wide at the lower end, about three times as long as wide, shouldering gradually to the shaft. On the western coast, paddle blades are somewhat more definitely pointed. In general, the paddle blades used in Malaya, Burma, China, and New Zealand are nearly square ended and widest at the end, tapering gradually toward the shaft. Among the Polynesians a pointed paddle is common, but the widest part is ordinarily below the center of the blade. The double-bladed paddles of the Eskimo are, depending on locality, both pointed and nearly square ended. Other double-bladed paddles are lanceolate, oblong, elliptical, or spear-shaped. In contrast to the Polynesian paddles, the African paddle blade is widest at the upper part, being heart-shaped or spear-shaped with the greatest breadth nearest the shaft.

A very cursory glance over a collection of paddles shows that although there is a great variation in usage of paddles, nearly all canoemen can and do paddle in all positions. Certainly, the African canoeman who stands to paddle large freight and passenger canoes, sits to paddle his small family dugouts. When I am in the bow of a canoe running rapids, with a companion in the stern, I want to kneel well forward; when I am paddling alone for an all-day trip I like to use a seat in the stern; and in approaching a little stretch of fast water I often stand up in the canoe or sit on top of the end piece in order to see what is ahead. When caught in a sudden wind flurry in an open lake, I usually shift my position from the stern to a kneeling position near the center of the canoe, where I have better control of the craft.

A great many primitive canoes have been profoundly affected by the impact of European culture. This influence is shown in the use of sails where originally sails were never used, or in the adoption of types of sails completely or slightly foreign to the native users. Probably sails are not indigenous to the canoemen of North America. All early travelers report that the woodland Indians, from New England to Minnesota and to the Arctic, did not use sails until the white man showed some of them how to use them. It may be that they did on occasion cut a bushy spruce tree and stick it in the bow for a sail when they were going before the wind on

Fig. 7.
Marshall Islands outrigger canoe.

straight stretches of river. But they seldom ventured out into the open lake where they might have found a sail helpful. Whenever an Indian travels, he watches for game trails and the discarded remains of captured prey, which he may see along the shores and which he would miss if he paddled far out in the lake.

The Aleutian open skin-boat, the umiak, on the Bering Sea carried a grass mat sail, but we may conclude that the sail was copied from Russian traders. Similarly, some of the far northern Indian canoemen have used sails made out of caribou skin, but again we may suspect the influence of white men, probably whalers or traders.

Even on the sea coasts, where the Indians developed quite remarkable and seaworthy canoes, they did not apply the use of sails. Perhaps this is good proof of the general isolation of the Americans from the rest of the world, where sails have been used from Europe to Singapore, from Singapore to Japan northeastward, and from Singapore to the far south of the Pacific Ocean. (See Fig. 7.) The use of sails over this vast stretch of the old world is no doubt due to accultural influences; that is to say that one race has borrowed from another.

When we look to South America we find remarkable sailing craft in Brazil, Peru, and on the great Lake Titicaca in Bolivia. On this lake, about two and one-half miles above sea level, there still persists a curious type of canoe made of bundles of reed tied together and fashioned into a boat shape. These curious and clumsy craft have had sails since before the days of white men. Not only did these people invent a sail but also a method of mounting it. They use a double mast, in the shape of shears, lashed together at the crossing. The sails were made of the same reeds as those from which the boat was made, and are tied side by side into a large mat. (See Fig. 8.)

The nearest other sailing crafts known in the new world were found on the coast of Peru. The astonished Spaniards saw natives sailing large rafts made of buoyant logs lashed side by side, supporting a little awning for the shelter of the crew. Not only sails but leeboards were used by these savages, with which they

Fig. 8.
Guayaguil jangada,
west coast of South America.

caused their boats to luff or fall away before the wind, depending on whether they thrust them down between the logs fore or aft of the craft. These craft also used the shear-shaped mast for support of the sail. As far as I am able to learn, those Indians on the interior of Lake Titicaca and those on the open sea coasts of Peru and Brazil were the only aborigines in all the Americas to have used sails before the white man came.

None of the Arctic kayak or umiak men in olden days used sails. They probably found no need for them. Almost all the tribes lived around the Arctic Ocean, in occasional contact with one another, and around Lapland in frequent contact with the western Europeans, especially the British and Scandinavians who certainly used sails. Around the Bering Sea they were in touch with Russian sailors long before the United States took over Alaska. Nevertheless, these sea hunters did not transmit the idea of sails to one another or to the Indians of North America. Long ago there was direct trade between the Eskimos and other natives of Alaska and the Asiatics, trading reindeer products from the Russian side for beaver skins from the American side. Neither of the groups used sails; consequently, sails could not have been transmitted to South America from the Arctic.

The Africans seem to have avoided sails but, like the woodland Indians, did occasionally put up a great frond of the raffia palm for a sail; at least that is reported from the coast of Fernando Poo. This again is by no means free from suspicion of the white man's influence. Although the Arabs penetrated Africa from the north and east, and are themselves famous sailors, they do not seem to have induced the Africans to use sails even on the great inland seas, the various Nyangas, and Lake Tanganyika. African canoemen of the interior seem utterly indifferent to sailing. However, in Madagascar and on the eastern coast, Africans have adopted a modification of the Indonesian sailing outrigger canoes. In Asia, Australia, and Polynesia some of the finest, most daring, most ingenious, and most skillful canoemen of the world are found. In America and Africa canoes are articles of necessity and utility, but in the East and on the South Seas they are objects of art.

Fig. 9. Dinkas spearing fish on the Lol River in Sudan.

African Dugouts 4

The Pygmies of the Congo basin in Africa have no canoes. Perhaps they are afraid of being caught out on the open river and killed by neighboring normal sized tribes. However, other jungle folk, who also live near rivers and among trees of a splendid type for canoemaking, have no real canoes. Some of the Negro tribes, apparently, are satisfied with nothing more elaborate than papyrus rafts. On the other hand, nearly all the more progressive tribes have canoes, generally plain dugouts, and in some cases more elaborate, built-up canoes. Nearly everywhere there are rivers and large trees in Africa, and men for many centuries have made canoes.

Before the coming of European explorers to central Africa, the natives had no knowledge of the saw as a tool, and many of them never even used wedges to split wood into slabs. Whenever they needed a plank, they had to hew the timber until it was thin enough to make a slab. For this reason they could hardly build a boat out of planks, or make a built-up canoe. They scarcely got beyond the simple dugout. Of course they knew nothing of joinery; and when patches had to be made on damaged canoes or when cracks had to be stopped, they lashed on a wooden patch with bark strings, or they plugged the opening or cracks with native rubber, clay, or grease. They made rope out of fibers separated from bark soaked in water. Holes were drilled in the wooden patches and in the canoes with red-hot irons. Using these holes and bark rope, they sewed the patches on while the rope was still wet. After this kind of rope dries it draws up very tightly and holds the patches closely in place.

After a tree has been felled, partly hollowed out, and roughly chopped into shape, a great crowd of people assemble and help haul the canoe to the water. They put down small branches for rollers upon which they slide and pull the canoe. In many cases, trees are felled at a considerable distance from the nearest water, and the labor of pulling the dugout is great. Often suitable trees are not readily accessible or not obtainable. In many tribes the trees are claimed as the property of the chief or headman, and he may not be willing to give over the best trees available.

The cost of making a large dugout in central Africa in 1935 was about fifty dollars. This cost is so great, even when paid with food or labor, that ordinarily a whole family or clan combines to make a canoe. The members of the group have a proprietary interest in the finished craft, and each may transmit his interest in the canoe to his heirs when he dies. It is no wonder that canoes are kept in commission as long as possible, patched and battered though they be. Most of the canoes used on the Congo and its tributaries are made from wood which will last about twenty years. In spite of care, many canoes come to a premature end. Some of them meet accidents in fast waters and rocky rivers, others are attacked by rogue hippopotamuses who sometimes "hole" a canoe or even bite the end of a canoe right off. Others, of course, are upset and broken by the onslaughts of great crocodiles. In every case, the canoemen struggle with all the strength and skill at their command to protect their precious canoes.

Although the Ubena use large canoes in freighting rice down their rivers, every man also has his own small canoe which he uses when fishing. When a Ubena does not have good luck fishing he blames it on the supposed or actual misconduct of his wife during his absence. This is the custom of the country, and wives are often beaten. If the luck is good, domestic peace and happiness reign.

In shape, the African dugout is fairly distinctive. The dugouts of the Nile, the Niger, and the Congo rivers are for the most part bluntly pointed at each end, nearly all ending at the bow in a short, solid, protruding snout. Similarly, the canoes used at sea by the Kru boys of the Guinea Coast have little solid snouts at bow and stern. The Kru are the best known boatmen of all Africa; they frequently take employment on European ships. Frequently they may be seen more than a mile from shore in their canoes. The canoe is made of a hollowed log, pointed at each end, without sheer, and with very little stability. The Kru, however, appear to be quite comfortable and secure, sitting upon a stout pole set into the canoe for a seat.

Some of the African canoes have square ends at the stern. Many of the river canoes made from tapering logs are cut square across the widest end for the stern. Placing the stern at the widest end is contrary to the practice of all modern boatbuilders, who always shape the widest beam forward rather than aft. Nevertheless, the African canoe-makers cling inflexibly to their established custom in this connection. Certain tribes, such as the Yapongu on the upper Lomami River (eastern Congo) and others in Nigeria, always make a square stern on their canoes, whether the original log tapers or not.

South of the Congo basin, in Angola, many canoes are made so large and steady that men can sit upon the gunwales without an upset. These canoes are characterized by flat bottoms, over-arching sides, slightly pointed ends, and no ornamentation except a little down-curved nose in place of the usually stubby snout.

In Madagascar, off the east coast of Africa, although the dugouts are essentially similar to those of the African mainland, the little solid snouts at bow and stern turn upwards and serve as mooring posts. Still farther north, along the east coast of Africa, the dugouts of the pearl divers on the Red Sea have no snouts at all. These canoes are used in the actual pearl fishing by two or three men put off from large sailing dhows. The canoes are simple dugouts with pointed ends, no sheer, and so solid at the ends that a hole is commonly bored through the bow or stern, from side to side in the solid wood, for the attachment of the anchor sheet or towline.

The dugout canoes of the upper Nile are so crude that some of them are not even pointed at the ends. They are nothing but the square ended, hollowed trunk of a palm tree that soon rots. Others, made out of solid wood, are pointed at the bow, rounded at the stern, but not otherwise shaped. As a consequence, these canoes are extremely apt to tip, very slow, and heavy to handle; and they require a marvelous sense of balance for their operation. The

Dinkas use such canoes for transporting their field produce to market and for fishing. It is to be expected that these canoes, like any other carelessly cleaned fishing boats, might have a strong odor; I am reliably informed that "they stink to high heaven."

When the men go fishing they use a very long spear, from nine to ten feet long. The spearman stands far in the bow, while his companion paddles in the stern. It is wonderful that anyone can stand and keep his balance in such a cranky craft, but that anyone can stand and manipulate a ten-foot spear on so quick a quarry as a fish is scarcely credible. As a concession to the possibility of capsizing, the paddler sits down. The spearman has the least possible advantage in steadying himself for the spear thrust, but a false or clumsy move by either man means an upset. However, many of the Dinka tribesmen do this regularly and would be surprised if anyone told them that it is rather a clever feat to spear a fish from a round bottomed, hollow log. (See Fig. 9.)

Dinka canoes are very slovenly in appearance; the outsides are left rough, the hollowing is irregular, the gunwale is uneven, the shaping is at minimum, the lines are very crude, and the sheer is small. The only ornament is a short snoutlike nose near the top of the bow. In spite of this, the Dinkas and their neighbors, the Shilluks,[1] use their crude dugouts with satisfaction and never abandon an old canoe until it is patched and cobbled beyond the toleration of their most tolerant good natures.

The Shilluks make large heavy canoes with thick planked gunwales and a high sheered, flaring bow in order to hunt hippopotamuses. (See Fig. 10.) They hunt the hippopotamus for revenge and for meat. Like most other African tribes the Shilluks are fond of meat but cannot always get it. Although they raise cattle, it is contrary to their customs to eat beef except when an animal dies. Since they may not butcher their cattle, they hunt the hippopotamus. This meat is so important a part of their diet that the king of the Shilluks claims as his royal prerogative a part of every hippopotamus killed.

---

[1] For information about Dinka and Shilluk canoemen, I am indebted to Dr. George Grabham and Mr. E. H. Nightingale, residents of the Sudan.

Fig. 10. Hippopotamus canoe and men of the Shilluk tribe.

Hunters paddle up to a chosen hippopotamus and spear him with a harpoon. (See Fig. 11.) If the hippopotamus charges the canoe, the men throw overboard a heavy wooden float to which the harpoon is attached and then jump out and swim ashore. The float betrays the position of the wounded animal, and the hunters pursue it in other canoes and finally kill it with long spears. If the harpooned hippopotamus plunges away down the river in the hope of swift escape, he often tows the canoe and the men with him. The high flaring bow on the heavy canoe plows through the water, throwing waves to both sides, while the canoemen crouch at the gunwale, peering through the spray, spears in hands. The great beast usually becomes exhausted in his frantic rush and is killed by the men in the canoe. Such a hunt is frightfully dangerous, but the hunters are generally successful. Around some African villages there are scores of hippopotamus skulls which are used by the natives as public benches. Old men sit upon these trophies boasting of their former prowess or gravely settling the affairs of the nation — or else they just sit.

Fig. 11. Ngombe hippopotamus hunter, northern Congo, with harpoon.

All the early travelers and missionaries in the Congo complained about the hippopotamuses and the crocodiles. As late as 1930, Towegale Kiwanga, an African author, lost all his belongings and a manuscript when his canoe was attacked by a hippopotamus on the Ulindi River. Nowadays crocodiles rarely attack canoes; but once a canoe is overturned and the men thrown into the water, the crocodiles are dangerous indeed. Crocodiles have learned the fear of gunshot. In Madagascar where crocodiles are numerous, the natives shout, beat pans, and slap the water vigorously in order to drive the crocodiles away.

Probably the greatest hazard the African canoeman has to face is that of human enemies. Not only has slavery been a common institution among African tribes, but slavery has been closely connected with cannibalism. When beans, millet, and kaffir fail, when fish are not obtainable, and when game is scarce, then men eat other men. When a war party raided a village the purpose was to secure all the villagers, dead or alive. If caught alive they were kept or sold to dealers for slaves, and if dead they were in many cases eaten.

For raiding and slave trading, many large dugouts used to be employed on the Congo and other large rivers. (See Fig. 12.) Much of the local warfare depended upon control of adequate watercraft, and large dugouts were often stout enough to stop a musket ball quite effectively. Many an African yarn features the fights between rival groups, each blazing away at heads bobbing above the solid gunwale of a dugout. If I had to choose between men, hippopotamuses, and crocodiles, I should welcome the risks of all the beasts of the jungle rather than those arising from the old-time African slaver.

Fig. 12. Dugout and paddlers on the Lualaba River.

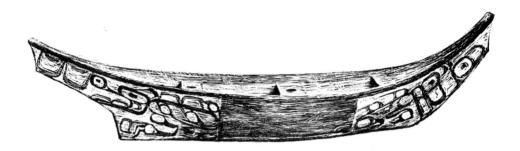

Fig. 13. Dugout canoe of Haida Indians.

5 American Dugout Canoes

All over North and South America dugouts vary in design and usage according to the trees found in the different localities. Excellent dugouts were made either of the western pine or the giant cedar by the northern Pacific coastal tribes. Both of these trees yield wood that is apt to split, requiring great care to prevent serious damage or even catastrophe to a canoe made from their timbers. Even the heat of the sun tends to make the canoes split along the grain of the wood in summer, unless they are left in the water or covered with brush. This tendency of the wood to split also affects the manner and technique by which canoes are cut out.

The Indians had to chop down one of the great trees with a stone axe. The outside was then roughly hewn to shape, and the interior chipped out by the very slow and laborious process of chipping with small adzes. A small adz is less likely to cause a serious split than a large heavy tool. The Indians used to burn holes in the hull and then chip out the wood between holes, the work of many months; then the sides of the dugout had to be spread in order to increase the beam. This was done by filling the cavity with water and heating the water by dropping in hot stones.

At the same time small fires were built along each side of the canoe near enough to heat the wood on the outside without charring it. The heat on the outside shrank the wood, and the soaking with hot water on the inside caused it to expand. The spread was made permanent by the insertion of crosspieces, as thwarts or seats, which were wedged or hammered into place during the process of spreading. How tremendous is the labor involved in making such canoes may be imagined when the size of one sample in the United States National Museum is considered.

This canoe is fifty-nine feet long; eight feet in beam; seven feet, three inches high at the bow; five feet, three inches in the stern; and three feet, seven inches amidship. Such an enormous canoe must have required a whole cairn of heated rocks to warm its bath. Even at that, the width of this canoe was probably not increased more than twelve inches by the hot water treatment. The original tree from which it was made must have been more than seven feet in diameter for a height of over forty feet, and very little less than that for about twenty feet more. Of course this is a monster; most of the canoes were much smaller.

The paddles were usually made of yellow cypress. They had crosspiece handles, often were carefully carved or painted in patterns similar to the characteristic totem figures used on the canoes, and had sharply lanceolate blades in contrast to the wide-ended paddles of the woodland Indians of the interior.

In southeastern Alaska the average size of canoes ranges between fifteen and twenty feet in length, carrying eight to ten natives with their baggage. One having a length of from thirty to thirty-five feet carries as many men. Smaller canoes about twelve feet long are made for the use of only one man. One of these little canoes is easily launched and handled, but the larger canoes require many men to launch them or to haul them ashore.

The dugouts of the northwestern coast of America had a distinctive shape. (See Fig. 13.) They all had a high bow and stern except for the small one-man boat. The bow and stern had a long overhang, and in many cases there was a vertical cutwater at the base of the overhanging, beaklike bow. In some cases they were made with flat bottoms, but otherwise rounded. They were made open for their full length, and the Indians relied on bailing to keep down the in-splashing water. The canoes were often highly ornamented on the outside with carved figures of sea lions, bears, seagulls, and other totems. They were painted or charred black on the outside, and decorated with a white or red stripe under the gunwale. The inside was smeared with yellow ocher, or in many cases left uncolored, ornamented only by the fine chipped surface produced by the adz.

Each canoe was carved out of a single log, with the exception that the bow and stern pieces were built upon the main part. These additional pieces were carved out of cedar and fitted to the canoe hull with wooden pins and rawhide lashings. As the green hides dried they shrank and bound the attached pieces still more tightly to the hull. The outside of the hull was very carefully shaped into fine lines, with the purpose of increasing seaworthiness and speed.

The early explorers spoke glowingly of the superb seamanship of these natives and of the excellence of their canoes. When Lewis and Clark first saw these canoes in 1805, they noted that they were considerably more seaworthy than white men's boats of equal size; and Captain Cook admitted that a crew of savages in a canoe could easily outdistance his own crew in their longboat.

For use on the tidal rivers and in sheltered bays from southern British Columbia to northern California, the western Indians used a much smaller canoe. For spearing salmon, which has been a staple of their diet for many generations, the Indians devised a shovel-ended canoe of rather clumsy but very steady qualities. An Indian could stand at the end of this canoe and spear a salmon under the very prow. These Indians never were in the habit of portaging their canoes.

From northern California to the coasts of Mexico the Indians used a dalca, a canoe shaped like a half-moon and made of several planks lashed together.

In the interior woodlands of North America, dugouts were made of light woods like basswood, white pine, or cedar. In shape the dugouts are much like those of the birchbark canoes, but they are heavier. On the eastern coast the Pequot Indians used to make a square-ended canoe out of Connecticut white pine. It was crude and clumsy, but it served their purpose in travel, fishing, and hunting on the lower tidal rivers of their country before they were driven out by the white settlers. In Virginia and farther south, the cypress offers a splendid medium for the making of canoes.

Today white men make dugouts after the old Indian design, using their own steel tools. Dr. Neil E. Stevens, the botanist, once told me that he used to own a much beloved dugout in North Carolina. It was made by local white residents out of southern cypress. Their practice is first to block out the canoe as to outside form by rough hewing from the log. Then they finish the shaping quite carefully, true to the final form that they intend to produce. With a brace and bit they drill holes into the bottom of the log two inches deep, and plug the holes with a different wood. They drill holes only one inch deep into the sides, and plug them likewise. Then they proceed to chop and hew out the interior until they have cut down to the surface outlined by the ends of the wooden plugs; thus they shape the interior to fit the outside with no real danger of cutting through the precious shell. When the canoe is soaked in water, the foreign wooden plugs swell up perfectly watertight. The canoe is made with both ends pointed, with a flat bottom amidship, and with no keel. It weighs between two hundred and two hundred and fifty pounds, it is from fifteen to seventeen feet long, about twenty inches in the beam, and about fifteen inches deep. The bow and stern are not sharply pointed, each being about five inches wide at the top.

On Chesapeake Bay one may see a more complicated form of dugout made of two logs joined together, side by side, to increase the beam, and with stern and gunwale built up to increase its seaworthiness. It is interesting to find that dugouts of aboriginal type are still made by white man blessed with modern tools.

In old Acadia, a large dugout, hewn from a white pine log, with broad sloping bow and square stern, is called a pirogue. In the southern part of the United States the pirogue is made of cypress, because, of course, there is no better native material. Annual regattas are held in Louisiana to decide the pirogue championship. A single paddler in 1939 made the astonishing time of over six and one-half miles an hour for a course of nearly five miles.

In the same district we find the frog hunter's canoe. It is very small, about eight feet long, twenty inches wide, and not over eight inches deep. The hunter catches his prey alive and unharmed by means of a spring trap at the end of a ten-foot pole. He works by day or night, using a miner's acetylene lamp in his cap at night. His canoe is made from southern cedar or cypress. It is very light, but easily tipped and barely large enough to float the trapper and his sack of frogs. It draws very little water and glides with ease through lily pads and over shallow muddy stretches of swamp. The frog catcher operates only in sheltered waters; his tiny craft would be useless in waves.

In contrast to the general use of soft, light woods for dugouts, the hardwood forests of the Middle West provide a black

Fig. 14. Seminole Indian dugout.

Fig. 15. Tarascan Indian dugouts with dip nets.

walnut dugout found on the Arkansas River. This hardwood is in striking contrast to the basswood, pine, cedar, and cypress used elsewhere, and probably owes its use to lack of a more easily worked, more buoyant, and equally sturdy, local competitor.

In the Everglades of Florida the Seminole Indians maintain in active use a native type of dugout canoe (see Fig. 14), which has been given up by the aborigines nearly everywhere else in North America except along the coasts of British Columbia and Alaska. In Mexico and Central America the dugout holds its place wherever there are lakes containing fish, or streams useful as routes of travel.

The Tarascan fishermen in Mexico go out in groups of six with wide spreading dip nets after small whitefish. (See Fig. 15.) The nets have a width of about sixteen feet and a height of eight feet. They are put into the water and raised again by means of a central pole. The fishermen take their places in a circle, with their nets submerged in vertical position. Paddling with one hand, resting the end of the handle against the shoulder, and holding the net with the other hand, they close the circle, driving a school of fish within the ring of their nets. At a signal, all sweep their nets up toward the surface and enclose the fish, dumping them into their narrow dugouts. Considering that the canoes are very small, just wide enough to permit a man to sit in them, wider at the bottom than at the top, and accordingly apt to tip, great skill is required to manipulate the nets without an upset. To use a paddle with only one hand and arm requires more than ordinary skill with a canoe. The Tarascans use this method of paddling every time they go out for whitefish. From the little canoes they transfer the catch into larger dugouts which follow the crew. The larger craft, with low, square sterns and high, pointed bows, are too unwieldy for use with the dip nets.

For ordinary travel the Tarascans use a larger type of dugout. On market days and fiestas, people gather from all around Lake Patzcuaro. It is a memorable sight to look over the azure lake, unruffled by breezes in the early morning, reflecting emerald foliage and the snowy peaks of the high surrounding mountains, as canoe after canoe comes gliding along, men in the stern and women forward. All these Mexican Indians use dugouts with peculiar flaring sides, widest at the bilge. Both men and women use a paddle with straight handle and circular blade, and sit down to paddle.

On the large inland lakes of Guatemala, men stand to paddle their canoes. They are made with high, built-up sides to keep out the waves on the stormy lakes, and only by standing can the men reach over the sides to paddle. In the rivers, however, the men sit down to paddle long, low dugouts, and on fast shallow waters they make use of punt-shaped canoes which they drive along with poles.

Seagoing canoes of great size and considerable variety are used on both the Atlantic and Pacific coasts of Central America. In general, they are hewn with thin sides and taper toward narrow, square ends. Very sturdy, large canoes are made out of mahogany or other hardwoods. Some of these are over thirty feet long, three feet in the beam, and so deep that men have to stand to paddle. They are so steady that a man may sit on the gunwale without overturning the canoe. Such stability is dependent upon the weight of the bottom of the canoe and upon its great length. Many seagoing, large canoes may be seen on the northern and western coasts of Colombia, where it is not uncommon, in order to add buoyancy to heavily loaded canoes, to lash alongside one or more balsa logs which act as huge floats to prevent the canoe from sinking too deep.

In many river villages of Colombia the family canoe is kept in water under the house which is built partly upon piles. In such villages the canoe is as much a part of family life as is the automobile in the United States.

In the interior of Ecuador, the Cayapa Indians make a dugout canoe which is among the most beautiful in the world. The natives lavish great care and labor upon their canoes. Good trees suitable for the making of fine canoes are so rare that it often requires an extensive hunt to find one. The Indian sets out through the forest, scrambling up steep slopes, through the tropical thickets garnished with stinging and scratching stems, often hacking his way through the liana-hung vegetation, paying no heed to flowers which would be priceless to the orchid hunters, but ever alert to find a suitable tree. When the tree found is two or three miles away from the nearest stream, the man and his family move into the bush and camp on the spot until the tree is felled. Then the log is left for several weeks, in order that it may dry out.

It is likely to require the help of ten or more people to transport the log to the home of the owner. There, beside his hut, the Indian will work away, week after week, hollowing out the canoe with an adz. To prevent the wood from splitting, he never makes large cuts and chops always across the grain. He chips the sides down to a thickness of about three-quarters of an inch, leaving the bottom between two and three inches thick. The ends are carefully shaped into slender and graceful forms. They turn up slightly and terminate in a square-ended extremity beyond a slender constriction. This is a highly artistic variation of the terminal platform at bow and stern which is characteristic of all the dugouts of South America.

After the Cayapa dugout is shaped and finished with the carving tools, it is coated with a preparation of beeswax, boiled down to suitable hardness. It is then painted with geometrical designs, some of considerable complexity. The beeswax finish gives the canoes an almost frictionless surface, which, added to the properties of their fine lines, must make them very speedy and easy to handle. The canoes are family conveyances; and men, women, and children all paddle. The women sit down to paddle, using short paddles with a crosspiece or an enlarged end on the paddle handle. They use the North American hold. The men stand to paddle and use long paddles with ornamented ends, over which they never place the hand in paddling. They use the hands-in-opposition grip. The men paddle in the stern and the women in the bow, exactly opposite from the practice of the North American Indian. Inasmuch as the Ecuador Indian stands to paddle, it is not necessary for him to be in the bow in order to see ahead, whereas the North American brave, who kneels to paddle, stays in the bow where he can see ahead much better than if he were kneeling in the stern.

Using his long-handled paddle, about a foot longer than a man is tall, the Cayapa man stands with one foot on the bottom of the canoe, and the other on the gunwale opposite from the side on which he paddles. The foothold helps him to keep his balance. He uses a large, family-size canoe, about thirty feet long, for traveling, and a much smaller canoe for general use. In both small and large canoes the man uses the same long-handled paddle, although the character of the stroke must be quite different. He sits to paddle in the small canoe. The man's paddle stroke is regular and relatively long; the woman uses short, jerky strokes in series of three or four with a rest between. There is no coordination, therefore, between the paddle strokes of a man in the stern of the canoe and those of the woman sitting forward.

The great interior of Brazil, the valley of the Amazon, and the surrounding territories of Guiana, Venezuela, Bolivia, Peru, Ecuador, and Colombia are the homes of numerous tribes using dugout canoes. They are either simple dugouts, or partly built up, with side boards and end pieces. The simple dugouts are known as the "Amazonian *uba*." One type is called the "duckbill *uba*," because it has a spoon-shaped, rounded, but not pointed, bow. It has either a square, overhanging stern, or a square, raking stern.

Another is called the "U-bow *uba*," because it is made from a hollowed-out trough only partly closed at each end. The center of the trough is spread open by means of heat treatment, is held open with thwarts, and the ends are raised high enough to prevent the water from running into the trough. Each of the *ubas* has square ends.

Among the Amazon tribes there seem to be no truly pointed canoes. The overhanging and spoon-shaped ends have certain advantages that pointed canoes lack: for instance, the flat overhang adds stability and buoyancy in rough or rapid waters. However, their canoe shapes are less likely to be the result of deliberate plan than to be the result of natural derivation from an original square- or round-ended log.

The people are of very primitive culture, and their canoes cannot be much in advance of their cultural development. They have been described as "folk without God" and "people without shame." They seem to have no religion and no personal modesty.

One of the explorers relates that he gave the women of one tribe pieces of red calico to be used as skirts. However, the women wore the skirts only because they found that the explorer was cross when they did not. The women protested that they did not wear the skirts in the sunshine because the light faded the color; they did not wear them in the rain because that made them wet; they did not wear them in the bush because the bushes tore them; and, of course, it was absurd to wear them inside their huts; therefore, why wear them at all?

Amazonian Indians do not appear to use the same method of paddling to maintain a straight course as the North American Indians do. North Americans paddle almost always on one side only, except for special emergencies, but the Amazonians appear to change frequently from side to side.

In the Gran Chaco of northern Argentina, dugout canoes on the Parana River are found with overhanging bow and stern, and with the bow slightly upturned. They differ from the Amazonian "duckbill *uba*" in having both bow and stern spoon-shaped, neither end being square. The Amazon and Parana canoemen harpoon huge fresh water fish and manatee; and before the days of firearms, they did not hesitate to attack the jaguar with their bows and arrows. They are men of great courage and make marvelous canoemen, especially adroit in running rapids, for which their canoes are very well suited.

In Argentina it is notable that the Indians stand to paddle their canoes on the large rivers. Perhaps this is because their canoes are made deep enough to outride rough waters. Certainly the Indians on the Atlantic Coast of Uruguay stand to paddle their seagoing, high-sided canoes. On the Pacific Coast, the Peruvian Indians also stand to paddle their dugout canoes.

In Peru the canoemen are handicapped for lack of local canoe timber. They have long been accustomed to import hardwood logs from Ecuador for the making of canoes. Those made of guachipilin wood are so durable that they last a hundred years with proper care. By contrast a balsa log lasts only a few months before it becomes waterlogged and sinks.

In the woodlands of Dutch Guiana live the Djukas, descendants of escaped Negro slaves, who threw off the bondage of their white owners and established their independence. The original slaves reverted to a life of savagery in a tropical forest much like that of their native Africa. Without borrowing from the local Indian cultures, they made their own adaptations to their new environment. One of their most notable cultural achievements is the evolution of a splendid canoe.

The Djuka canoe is a dugout, hollowed out all the way to the very point of bow and stern. Unlike the Indian canoes, it has sharply pointed and upturned ends. In size the canoes vary from eight to forty feet in length. A forty-foot canoe with four paddlers will carry safely a thousand pounds of baggage.

Djuka canoemen are sometimes hired to carry Europeans up and down the South American rivers. These rivers are infested with swarms of a bloodthirsty fish, called piranha. These fish look something like a very large-mouthed bass with long, pointed, slashing teeth. A school of piranha will tear a human being to pieces. In spite of the piranha the Djuka canoemen never hesitate to jump into the water to protect their canoe and passengers from a smash-up or swamping, although they seldom bathe in the rivers for fear of these fish. A Djuka will run any risk to protect his passengers, knowing that he would be ostracized by his people if he should ever suffer so terrible a disgrace as to lose the life of a passenger.

The Djukas usually make their canoes out of the Guiana cedar, a soft lightwood, known all over the world as that from which cigar boxes are made. The canoe-makers leave the Guiana cedar to season for several months after being felled, before they hollow out the center and hew down the sides to perpendicular walls, leaving a round bottom. They fill the wooden shell with water, heat it with hot rocks, soften the wood thoroughly, and force it into the desired shape. The ends and seats of the canoes are often decorated with intricate carvings. In spite of their very delicate lines and their undoubted buoyancy, these canoes are actually quite heavy. It requires the strength of several men to lift an empty canoe over the rocks in rapids, or to pull it up onto the muddy banks of the tidal rivers.

As with the canoe, so with the paddle: the Djuka paddle is quite different in shape from those of the Indians. It is broad at the lower end, tapering gradually to the lower-hand hold; whereas the Indian paddles usually have long blades of equal width throughout, or else they are broad in the middle and tapering both ways. The Djuka paddles are hewn from a hardwood plank, and frequently are decorated with fine carvings. A man's paddle is as long as he is tall, but a woman's paddle is much shorter.

The Djuka use poles on rapid rivers. The canoeman uses a ten-foot pole and pushes his canoe upstream by walking along the gunwale, pushing on the pole until he reaches the stern, when he returns to the bow and plants his pole again. This is much like the practice of the African tribes on the northeastern tributaries of the Congo. It is said that the Djuka canoes show strong West African affinities, but they are certainly distinctive in many respects. For sheer beauty of form and excellence of workmanship they are approached by no dugouts of South America or of Africa, except perhaps by those of the Cayapa Indians in Ecuador.

Fig. 16. Shilluk papyrus canoe, upper Nile.

6 Reed Canoes

Moses in the bulrushes is a tale often parodied and distorted, but, for all that, it remains one of the most beautiful and touching tales of ancient literature. Bulrushes are pithy, porous, and very buoyant plants, quite competent, when bound into a raft, to support a child's weight. Perhaps Moses' mother had seen large vessels made of bulrushes in ancient Egypt. Isaiah XVIII:1-2 says, "Woe to the land shadowing with wings, which is beyond the rivers of Ethiopia: That sendeth ambassadors by the sea, even in vessels of bulrushes upon the waters."

Even though Isaiah was not quite literal in his allusion to bulrushes as shipbuilding material, undoubtedly similar materials were used for vessels in ancient times. Certainly in recent times bulrushes have been used for seagoing canoes by the Maoris of

New Zealand. About eighteen hundred years after the days of Moses, Pliny recorded quite specifically that the Egyptians made ships out of papyrus and bulrushes. Pliny thought so well of the Egyptians' craft that he considered them to have been the original models of all ships.

Papyrus, a sort of pithy reed, grows in shallow waters all over central and northern Africa. It is well known as the plant from which the ancients made their paper, and as a boat-making material it is so old that it was legendary even in the days of Plutarch, a contemporary of Pliny, who wrote about 70 A.D. The story Plutarch wrote was an ancient tale of the mythological Isis who traveled through the swampy wastes, endlessly searching for the body of Osiris; she used a papyrus raft, bound about with papyrus cords, and tied together into the shape of a canoe, both ends pointed and turned up.

The ancient papyrus boats were usually of this shape. In a mosaic pavement recently unearthed at Praeneste, near Rome, there are illustrations of papyrus float-craft with both ends pointed and upturned. Figure 16 shows men of the Shilluk tribe paddling a

Fig. 17. Shilluk canoe, upturned only at bow.

large papyrus canoe, pointed at each end and with both ends turned up. It is scarcely believable that the use of so primitive a craft, made out of river reeds, should persist for over five thousand years. Nevertheless, this type of canoe is probably more ancient than the oldest pyramid.

The commonest form of modern papyrus canoe is upturned only in the bow. (See Fig. 17.) This type of craft is reported from Lake Moeris in lower Egypt, where it is propelled only with poles, the lake being very shallow. The Shilluks are said to use the "pole-like stems of a bush which grows in the water, composed of wood as light as cork," according to a description of these craft in 1869. They are still being used by the same tribe. They are brought to a long tapering point at the bow and have a sharply rounded stern. The stems of the plant are bound into a sort of rough gunwale on each side, leaving a shallow hollow in between. The craft is supported by the buoyancy of the stalks and not by the displacement of the hollow shape as in the case of an ordinary canoe. This model

Fig. 18. Large reed canoe on Lake Tana.

has neither end upturned, and is propelled with a pole or a paddle. Some of these papyrus canoes are large enough to carry large loads of native produce and other belongings. Even though they are slow and awkward to pole or paddle along, they are desirable in Africa, where all too often freight has to be carried on human heads and shoulders.

Larger reed canoes are used on Lake Tana, at the headwaters of the Blue Nile in Ethiopia. (See Fig. 18.) On islands in this lake are some of the most ancient Christian monasteries in existence. The great antiquity of their records and their remote and secluded position make them of special interest to scholars. Major R. E. Cheesman visited several of these monasteries in 1933, and the following is a sketch of the setting in his own words.

> I set out on a cruise around the lake, covering two hundred and twenty miles by water. There are no boats on Lake Tana; reed rafts called *tankwas*, propelled by Waito sailors, are the only means of transportation.
>
> This six weeks' journey around the lake by *tankwa* was an unending dream of delight; we never knew from day to day what surprise was in store on the next island, or round the next headland. We were always in magnificent scenery. In the foreground were the blue waters of the lake which rippled against the sides of the *tankwas* as the men bent to their bamboo poles, while beyond, the colours of the forest on the shore and mountains made an everchanging scene of exquisite beauty.[2]

Lake Chad is a shallow, swamp-lined inland sea between the headwaters of the Nile, Niger, and Congo rivers. Along the reedy shores, and on hidden, almost inaccessible islands live the Buduma tribe, known as pirates. These people use canoes fashioned out of bundles of reeds, large enough to carry twelve or more men and baggage, donkeys, etc. They are really rafts of great buoyancy, made in elongated shape, with one high-pointed end. They have low wide bulwarks, each fashioned from a long bolster-like roll of papyrus reeds lashed together. They are propelled with poles, twelve to fifteen feet long and not particularly straight. The men pick up any pole long enough to serve the purpose, and they use no sails and no paddles. The canoes gradually become watersoaked

---

[2] Major R. E. Cheesman, "The Highlands of Abyssinia," *Scottish Geographic,* 52 (1936), 8.

and sink in a few weeks, but each man seems to own his own canoe and to make a new one from the reeds along the shore whenever he needs one.

In lower Senegambia and along the coast of Guinea, the natives use a sort of small grass boat. This craft is not much more than two cigar-shaped bundles of rushes firmly bound together. A wide bulwark of additional rolls of rushes makes a canoe-shaped craft out of the contrivance. Until they become watersoaked they are practically unsinkable. Both ends are pointed and slightly turned up in archaic style. The craft appears to be a throwback to the most ancient type of reed boat used by the Egyptians. They are used on the rivers and even as surfboats on the sea.

Still farther south we find reed rafts on the Kasai River, among very backward tribes in the lower Congo basin. These people, the Bakwese, have no canoes, even though good canoe timber is all about them. Other Congo Africans who use similar rafts are the Lomami on the upper Congo, but these people also have good dugout canoes.

Reed canoes are found in Africa scattered over an area about twenty-five hundred miles from north to south, and about three thousand miles from east to west, by far the largest area in the world within which the reed canoe is known to be in use.

On the lower waters of the Indus River in northwestern India fishermen take bundles of buoyant shola stems, sling them under their armpits, and float down the river with their fishing nets. Having caught their fish, they trudge off to sell them, carry-

Fig. 19. Tule grass balsa on Pyramid Lake.

Fig. 20. Little balsas on Lake Titicaca.

ing their floats with them. But when sudden floods spread terror and distress over the countryside, the natives hurriedly cut green bundles of the cattail rushes which grow very large in the valley of the Indus. Men put their wives and children on rafts made of these bundles, called tirhos, which are really emergency lifeboats. These floats indicate the evolution from a very simple float to a raft, and thence to a canoe-shaped boat.

On the upper Indus the natives make floats out of animal skins stuffed with buoyant stalks and straw. These are used as ferries, and combined they make a raft on which heavy produce is floated down the rivers. Here we see a break in the development of boat design; one line develops the mashak and skin-covered canoe, and the other the tankwa or reed canoe.

Long ago the North American Indians on Pyramid Lake, Nevada, began to use a canoe made out of tule grass — a very tubby, crude type of craft sixteen or seventeen feet long, and over five feet in the beam. (See Fig. 19.) Each end was roughly pointed but not upturned, in striking contrast to the half-moon shape of the reed craft found in South America. A more shapely grass canoe comes from Thuron Island, the Gulf of California, in Mexico. It is almost almond shaped, with high and very tapering ends. It is made of one bundle in the middle about ten feet long, and two more, one on each side, twenty-seven feet long, with a maximum width of about three feet. This craft was much more easily navigated than the short and squat balsa from Pyramid Lake.

The balsa is also found on Lake Titicaca in Bolivia. The general method of make-up and perhaps the original form of the balsa is indicated by what they call the *caballito* or little balsa. (See Fig. 20.) The *caballito* really consists merely of two spirally bound bundles lashed side by side. This makes an almond-shaped float, about eight feet long, just big enough to support one rider. *Caballitos* are still used occasionally in many Peruvian harbors, and in the old days they were common.

The largest number of balsas, however, are used on Lake Titicaca. This great lake is about 150 miles long and sixty miles wide. It stands at an elevation about two and one-half miles above sea level, near the heart of the Andes Mountains, and is subject to terrific squalls common in high mountains. The water is so cold that anyone swept overboard would soon perish from exposure. For these reasons canoemen on Lake Titicaca need a craft as nearly unsinkable as possible. The balsa cannot be sunk except by being in water until it becomes either rotten or waterlogged. For long trips, in the exposed parts of the lake, the natives use a large balsa made on a model considerably different from that of the *caballito*.

The large balsa is as much as twenty-five feet long, and it can carry several men or animals and a considerable cargo of freight. It is more than five feet wide, it rides about thirty inches above water level, and has a distinctly hollowed or boat-shaped interior. Although the sides are built up with bulwarks, the fundamental structure appears to be two very large bundles which taper sharply at the ends. These ends are turned upward side by side, in a definitely recurved shape, but are cut off at a level about ten inches above the line of the gunwale. The top of the gunwale is ordinarily nearly level in the case of the larger type of balsa, whereas in the smaller ones there is often a rise in the level of the gunwale amidship.

Fig. 21. Large balsa with sail.

No sails were used in North America before Europeans introduced them, but in South America they are of great antiquity. The sails used on Lake Titicaca watercraft are made of reeds tied side by side into a large mat and supported by slender yards at top and bottom. Each is suspended from a native type of mast unlike any other mast in the world. It is made of two sections, often composed of several sticks lashed together, for timber is scarce in that part of the high Andes. These are stepped into the bulwarks and lashed in place. They are crossed and lashed together near the top, and they are braced by a long three-pronged pole, the foot of which is stuck into the bulwark near the stern of the balsa, standing about forty-five degrees to the two masts. The sail is hoisted upon the windward side of the mast, which stiffens it and gives it support against too strong a breeze.

For a large balsa the mast is about ten feet tall; the sail head about nine feet; luff, three and one-half feet; and leach, twelve feet. (See Fig. 21.) The sails are always used as lugsails. Usually the balsas plough along with the sails set low and wide, which tends to prevent too serious a cant before the wind and helps to hold a course oblique to the wind. In other cases the sails are made high and narrow. This requires a particularly sound pair of masts, which are not usually available, and it increases the danger of a sudden wind wrecking the sailing tackle. The large balsas cannot respond quickly to a wind load as a very light craft can; consequently, the strain of a sudden breeze must be carried by the sails and masts until the craft has slowly gathered speed.

Australia and Tasmania have their representative rush and reed canoes. Although the Australian aborigines had other forms of watercraft, the old Tasmanians did not. In Tasmania the bundle-of-rushes type of craft seems to have made no further progress. It was made from strips of bark of the swamp tea-tree, which is soft, velvety, and very buoyant. The natives tied the bark into a canoe-shaped bundle, pointed and upturned at each end. It is said that they used to sit on this sort of float-canoe and paddle it with sticks or poles. Rather better floats and rafts made of reeds have been seen on the rivers of southeastern Australia. It is said that some of them were large and strong enough to carry several natives, their children, and their scanty belongings.

In New Zealand the early inhabitants, perhaps the aborigines more ancient than the Maoris, made canoes entirely of the bulrush. One of these vessels has been seen nearly sixty feet long, capable of holding as many persons. They were remarkably thick, formed entirely of rushes, except the thwarts, and resembled the model of a canoe in every particular.

The Maoris made a canoe built essentially of raupo leaves tied into bundles. This was as much a raft as a canoe, depending upon the buoyancy of the bundles of leaves for its capacity to float, and being very temporary in nature. Inasmuch as the Maoris have excellent dugout canoes, these raftlike craft are of much less practical importance than are the balsas among peoples who really depend upon them, but they must have been of great ceremonial importance to the Maoris.

The Chatham Islands, which lie over five hundred miles east of New Zealand, were inhabited by the Moriori.[3] These islands support no large trees. The largest bush they boast is the tree-shrub called akeake. The wood of this shrub is very durable and tough. However, it is never large enough to provide a dugout canoe, and yet it does provide small slabs about two feet long and six or eight inches wide, and poles long enough for gunwales. There is also abundance of a tough bramble called supplejack in New Zealand, which was used as a binder. They also had the matipo shrub to provide the various poles they needed. Very buoyant materials are provided by the so-called bull kelp from the sea, and the wild flax and ferns which grow on land. Using only these sticks and stalks, the Moriori made seagoing canoes which challenge the admiration of every educated man.

All the large Moriori canoes were made with tholepins built into the gunwales, against which the canoemen braced their paddle handles and rowed with their backs to the prow — a very unusual position for any of the natives of Australasia or of Oceania

---

[3] In 1901 the population in the Chatham Islands was little more than 400, only 31 were Morioris and 18 of these were half-breeds.

in general. The larger canoes, called *waka-pahii* were fifty feet long, eight feet in the beam, and five feet in depth, and must have required a large crew of paddlers.

The Moriori used the large canoes to hunt the blackfish, a small whale that swims in schools. These whales play a follow-the-leader game in the water, all leaping out of the water and diving beneath the surface, showing their shining black backs in unison as they keep time with their leader. The Moriori, hunting these schools of whales, drove the leader and the whole pack into the shallow waters where they butchered them wholesale. Women and children splashed out into the shallow water near shore, shouting and cheering, as the men in their great hunting canoes came rushing in after the blackfish, partly stranded in the reef. Daring spearmen and harpooners dashed in among the thrashing tails to drive home the deadly thrusts, while waves splashed high and the canoemen rowed in the open sea as close to the shore as they dared to come in their flimsy and rather unwieldy craft.

A good kill of blackfish meant food for all the Moriori for weeks to come, but without canoes to drive them into shore, the Moriori must have starved. Except for blackfish, seals, fish, and birds' eggs there was no food to be had on the scrub-covered islands.

People who had the ingenuity to make canoes out of leaves and sticks must have been extremely interesting in many other ways. One of their customs required the construction of a large low canoe made of fern stalks and flax, called *waka-ra*. In it were placed the wooden images of men, each with a paddle lashed to the hands. This craft was sent out to sea with a fair wind, as an offering to the god Rongotakuiti, who in response was supposed to send ashore shoals of seals and blackfish.

Fig. 22. Greenland kayaks.

Skin-Covered Canoes — Kayaks 7

The Eskimos made the best skin-covered canoes in the world, the best known of which is the kayak. This is a canoe made for a single occupant, and it is used so generally by the hunter of the family that it is called by the Eskimos the "man's boat." It requires a man to use it. The kayak is completely decked over except for a cockpit into which the canoeman ties himself. (See Fig. 22.) It is so light that a man can pick it up with one hand, and so buoyant that it floats like a cork on the water.

The kayak is used almost all around the Arctic seas. It was in use in Greenland by the skraellings, as the Norsemen called all savages, when Gunnbjörn discovered the land a century before Eric the Red tried to colonize it in 986 A.D. The kayak was the only small craft known in far northwestern Russia when Stephen Burroughs reached Vaigach in 1556, and it was found along the coast of Alaska and among the Aleutian Islands in 1778 by Captain James Cook, the explorer. Nobody knows how ancient the kayak is or who first invented it. It is one of the most wonderful crafts ever devised by man. It is built exclusively of materials at hand, and for the most part produced nowhere else except in the Arctic seas. It is the best small craft that has ever been invented for the hunting of Arctic animals and for use on the open Arctic seas.

The Eskimo sits on the bottom of the kayak — a position of considerable discomfort for the average white man — and is sometimes kept at sea by offshore winds for many hours at a time. So marvelous is the endurance of the Eskimo in hunting or paddling that on such occasions he makes fast his kayak to the lee of an ice cake and waits for the wind to change. There he has to stay, hour after hour, with no change of position possible. Occasionally he falls asleep in his seat, slumped over the harpoons and paddle on the front deck of his kayak. In many cases he faces the alternatives of drowning in a kayak smashed against ice floes or of starving to death. Most Eskimos cannot swim. Usually, however, the wind changes in time to let him get back to his igloo safely.

The Eskimo has to paddle in water that is nearly or actually at freezing temperature; to get wet and remain wet under such conditions is fatal. To keep himself dry, the Eskimo wears a hooded jacket made of waterproof seal gut. This garment fastens with drawstrings under his chin and around his face. It is lashed tightly around the cockpit, effectively keeping out any water that might splash on the canoeman and run down his clothing into the craft. Only the face and hands of the Eskimo get wet, and he protects his hands from the drip of water running along his paddle handle by two rings of ivory or bone placed just beyond his hand holds. Most of the water drips off the rings, instead of running up his sleeves

Fig. 23. Eskimos somersaulting in kayaks.

and making him cold and wet, a dangerous condition in that temperature.

The Greenland Eskimo uses a double-bladed paddle from six to seven feet long, with either lanceolate or square-ended blades, depending upon the locality. The canoe is about twelve feet long, and two feet wide, and having little keel it swerves from side to side with every stroke of the paddle. Consequently, when a kayak man is pursuing an animal into which he wants to plunge his harpoon, the last stroke of the paddle before he grasps his harpoon takes him on a swerving course at the very time he must make his strike. To remedy this inherent defect, the Eskimos of southern Greenland use a small wooden paddle as a rudder on such occasions, probably a local adaptation of the rudder which they have seen in whalers' boats.

The Eskimos are so clever in the use of the kayak, that the language has a word which means that a man is able to turn a side somersault with his canoe, passing entirely under water and righting himself and the kayak again without touching bottom. (See Fig. 23.) It is said that some can do this feat several times in succession like a tumbler turning over and over.

During the seventeenth and eighteenth centuries several kayaks were captured in Scottish waters, one having been taken near Aberdeen where it is still preserved. The men who handled those kayaks were almost certainly Samoyeds or Laplanders, probably related to those who once lived near the mouth of the Petchora River in northeastern Russia. For many years it was thought that these canoemen were Eskimos who had crossed the Atlantic from Greenland. However, examination of the wooden framework of the kayak in Aberdeen shows that it is made from wood which grows only in northeastern Russia.

As a matter of fact, the kayak seems to be in use all around the Arctic, varying in design from place to place. The Alaskan kayak has a vertical and square stern (see Fig. 24), whereas the Greenland kayak is pointed at both ends. The top of the Alaskan deck is ridged down the middle; the Greenland kayak usually has a flat deck. The bow and stern ends of the kayak are tipped with bone or ivory in Greenland but not in Alaska. The Greenland kayak has a bone cutwater keel under both bow and stern extremity, but the Alaskan has those parts unprotected. In many Alaskan models the gunwale rods protrude at the stern for a handhold, or else a pocket is made in the sealskin covering within which one may grasp the frame.

Fig. 24. Alaskan kayak.

Fig. 25. Labrador kayak.

In the Aleutian Islands, west of Alaska, the kayaks in some cases have a double-ended bow, one of which is upcurved and reversed like the head of a violin. Occasionally also the Aleutian canoeman uses a single-bladed paddle, almost unheard of among the Greenlanders.

The Labrador Eskimo uses a wider, longer, and heavier canoe than does the Greenlander. The long kayak makes it possible to use large paddles, from ten to twelve feet long. The stern of the kayak is wide and flat, and the bow raised and pointed. In former times they sometimes made the bow upcurved. Nowadays a kayak with both ends upturned (see Fig. 25) is used farther west, about the mouth of the Mackenzie and Red rivers. This type of kayak is not found nowadays either in Labrador or in Alaska.

The Greenland Eskimos, who still live much as they have always lived, and who still make their kayaks with the care and pride of the old master craftsman, today have the best kayaks and the best kayak men. Men who harpoon walruses and various kind of whales and who fight and kill polar bears in such frail little craft have to be skillful canoemen. Nevertheless, they have a marvelous canoe to handle. The kayak is so convenient for work in Arctic waters that it was used in preference to other craft for surveying work by the staff of the National Geographic Society Expedition to Alaska in 1919.

Kayaks of all types are built with a framework which is constructed independent of its covering. The upper part or deck of the frame is practically completed as the first stage in construction. Later a middle strip is fastened along the top, intersecting the opening or manhole.

Since the manhole supports not only the position of the canoeman, but also the attachment of harpoon lines, it must be braced solidly into place in relation to the framework of the whole canoe. After the framework of the top is complete, it is turned top downwards, and the ribs reshaped, fitted, and their ends inserted in the upper side pieces (thwarts) to which they are secured with wooden pins. The ribs are usually from two to six inches apart. Longitudinal strips are then attached to the sides, with a similar strip along the middle of the bottom.

Men make the framework of the kayak, and the women sew on the skin coverings. The covering is of sealskin sewn together and attached to the framework while wet. It stretches tightly as it dries and hardens into a set shape. The setting of sealskin after drying is so hard that when an Eskimo breaks a bone he wraps the limb in wet sealskin and lets it dry into a rigid splint.

In order that the covering can be sewn together and applied in a short space of time, it is done by a group of women working together. In the old days women had to do the sewing with ivory needles or with needles made out of hares' teeth which had been sharpened. Sewing was slow work and required several workers for each kayak.

The Eskimo women enjoy these occasions; they correspond to the old-fashioned sewing bees of American women. The Eskimo men complain that the speed of the needles never keeps up with the tongues, but they hang around, nevertheless, to share the gossip. Eskimo women have no inhibitions in conversation, and Eskimo men have no matrimonial jealousies; consequently, gossip at the kayak sewing party seldom leads to anything but good humor.

A special kind of skin boat, like the kayak, but made for the use of two or three canoemen, and having a corresponding number of cockpits, is called a bidarka.

At the present time the Aleutians and Alaskans are the only peoples known who use two- or three-hole bidarkas. Captain Cook in his famous voyages found natives of the Aleutian Islands using single-bladed paddles in two-hole bidarkas with a highly turned up bow and a low stern. At the same time, he found also the one-man kayak with its occupant, of course, using the double-ended paddle. However, in 1653 a two-hole bidarka was brought back by a Danish expedition from Arctic Russia. From this it is clear that the bidarkas formerly must have been more widely used than they are today, when they are used only throughout the Aleutian Islands. In making explorations in Alaskan waters in 1885, officers of the United States Revenue cutter "Corwen" used a three-hole bidarka especially made for their needs. It is now preserved in the United States National Museum at Washington. It is twenty-five feet long, thirty inches in the beam, and only ten inches in depth.

It is said that when Russian traders first began to traffic among the Aleutians, they were practically helpless in a kayak, and more of a danger than a help to an experienced native in a bidarka. Russian traders have never had a high reputation for seamanship. The Aleuts solved this problem by making a new kind of bidarka, one with three holes, so that the trader could ride as a passenger, while two natives, one ahead and the other astern provided the paddle power.

An open skin-covered boat, able to carry forty or fifty persons, is known as the umiak, or the "woman's boat." (See Fig. 26.) It is never decked over, it has a flat bottom and high sides, and it usually is built with transverse seats. The stern is not pointed and the bow is only slightly narrowed. The umiak is not used exclusively by women, just as the kayak, known as the "man's boat," is not used exclusively by men. As a rule women do not go out in

Fig. 26. Alaskan umiak, sealskin boat.

kayaks; when it is necessary for women to move with the family or village, they use the umiak. This is constructed over a framework made of wood or whalebone lashed and mortised together, and the walrus skin cover is lashed over the top of the gunwale and fastened to rawhide ribbands which run under the gunwales for that purpose. The hides are stretched by pulling on these lashings. During the winter the rawhide cover is taken off and folded away for the season.

Driftwood is the only wood available to the Eskimos for the construction of their canoes and weapons. The Christianized Eskimo who finds tree trunks trapped in the deep fjords of southeastern Greenland, brought there by the Gulf Stream, calls the drift "Noah's wood," and regards it is a residue of the Flood.

Captain Fridtjof Nansen records that the Eskimos of eastern Greenland journey to the Danish settlements on the West Coast for the sake of getting a supply of snuff — a journey which requires two years for the round trip. Such a journey around the ice-skirted edge of Greenland's fjords and precipitous cliffs is exposed all the way to the open Atlantic and is made up of short trips between resting places.

Every five or six years a clan may decide to conduct a walrus hunt. Using both kayaks and umiaks, the clan moves bag and baggage to some locality known to support walrus herds. They settle for the time in a place convenient for their walrus hunting, and when sufficient hides and ivory have been collected there they move on to another hunting ground. They use for hunt or travel whichever type of skin-covered canoe best suits their immediate purpose. Their nomadic life, the only life possible in an Arctic setting, is made much more convenient by the use of canoes in summer and dog sleds in winter.

Many men the world over need watercraft, but scarcely any race is so completely restricted as to available materials as are the Eskimos. The Eskimo canoemen may be taken as models of human adaptation to their environment.

Fig. 27. Coracle, yak-skin boat, on Tibetan stream.

Primitive Skin-Covered Crafts 8

To the plains Indian the horse was the natural means of conveyance, corresponding to the birchbark canoe of the woodland Indian. Horsemen had no need for a real canoe, and without necessity there seems to have been little invention beyond the bull-boat, a primitive coracle. It was merely a rough framework of willows, covered with buffalo hide, about five feet in diameter and eighteen inches deep. To make it go, the Indian used a short-handled paddle with which he reached out as far as he could and then pulled in straight toward the boat. It did not make very good speed, often drifted, and usually had to be towed back upstream by a horse. Unlike the canoes of other races the bull-boat was not used for fishing and hunting, but only to help the women bring in firewood or as a ferry.

In South America the only coracle comparable to the North American bull-boat is called a *pelota*. It is used in the Gran Chaco region of Paraguay, and in Argentina. It is described as a square boat or tub of leather. They say that the Indians in a *pelota* hold on to a horse's tail and let the animal tow them across the river. If no horse is available, the natives swim behind and push the *pelota* along. In a similar manner the Ethiopians wade and swim alongside their leather-covered round boats called "jendies." In these jendies they ferry passengers across the Blue Nile; animals have to swim.

Coracles made of wickerwork and covered outside with pitch are nowadays used often in Iraq, formerly Mesopotamia. Some of them are large enough to transport several tons of freight. Other coracles covered with skins are found on the waters of the Euphrates and Tigris, where they have been used with almost no important change in design since the days of antiquity. Herodotus reported that the country folk of Mesopotamia floated down the river in skin-covered boats carrying their produce to the large cities. When the owners had sold their goods, they disposed of all parts of their boats except the skin coverings. Those they packed on the backs of asses, and took home for use on later trips. Such boats still in use are known locally as gufa. They are propelled with paddles, but on account of their round shape they are not speedy or easy to handle. For one-way, downstream traffic, however, they are cheap and adequate.

The coracle covering changes from place to place with the domestic or wild animals locally available. In Tibet the shaggy yak or mountain ox is the mainstay of human life in the high Himalayas and provides a tremendously thick and tough hide. Coracles covered with yak skin (see Fig. 27) are used as ferries for crossing the headwaters of the Yangtze Kiang in Tibet and far western China. Water buffalo hide is used for similar round boats in many parts of India. Where there are no hides available, the people may use either a pitch covering as in Annam and around the Persian Gulf, or even pottery.

In Bengal where wood is scarce, no bitumen is obtainable,

and people are too poor to buy leather for covering basketwork, the village carpenter makes an earthenware pot large enough to float one person. This round pot is called a *tigari*. In order to make it go, without helplessly spinning around in one place, the native squats in the *tigari* and paddles with his hands on both sides at once. The *tigari* is a source of much amusement to the natives themselves; they stage *tigari* races at their village fairs and give prizes to the winners.

Until very recent times, coracles made of basketwork and covered with leather were used by both Scotch and Irish peasantry for salmon fishing. A coracle formerly used on the River Boyne, Ireland, is now in the United States National Museum. It is nearly six feet long, almost four feet wide, and seventeen inches deep. Julius Caesar said that the ancient Britons were painted savages and used skin-covered coracles. So did many other ancient people of northwestern Europe. Such a craft costs relatively little, will float in very shallow water, and is so light that a man can easily carry it home from the river.

In Ireland and Scotland the coracle was improved by an increase in length and conversion into a canoe. In Scotland these were called currach and were still in use in the sixteenth century. In Ireland they have continued in a somewhat revised form, the leather skin having been replaced by cheaper, painted canvas. On

Fig. 28. Irish curragh.

Fig. 29.
*Mashak, inflated pig or cow skin,
used on the Sutlej River
in the Himalayas.*

the southwest coast they are larger than in the north. It is the small northern craft which is called a curragh. (See Fig. 28.) In general, the curragh is a canoe with square stern and full round bow. A specimen in the United States National Museum is seven feet, ten inches long; three feet, four inches in the beam; and twenty-five inches deep.

The curragh is also remarkable for the fact that it is used with a curious form of oar. The oar end is square and not bladed; furthermore on each oar is attached a triangular piece of wood, pierced with a hole large enough to fit over a tholepin in the gunwale of the curragh. This is a canoe-like craft made for use with oars. The curragh floats lightly on the water and is extremely seaworthy. However, care must be taken to prevent its being thrown by waves against a rock or upon the beach, because it is of very frail construction.

The very simplest skin craft of all is a skin sewn up, inflated with air, and used as a bladder-like float. Such contrivances are common on the rivers of the Punjab in India where the skins of buffalo or sheep are used for the floats. The skins are so well sewn up and inflated that when ready for use they closely approximate the form of the living animal (see Fig. 29), and the natives ride them astride on the water, paddling with arms and legs. This requires considerable practice, for the craft is easily rolled over. A young English officer who tried to navigate one of these mashaks was nearly drowned.

Obviously a mashak is a poor substitute for a boat, unless several are lashed together to make a raft. Such contraptions have long been used in Assyria, Mesopotamia, central Asia, northern India, and China. In China three rows of four inflated sheep skins are attached to a framework, on which freight and passengers are carried. In India, buffalo skins are joined together with a sort of basket top as a deck. The usual name for all rafts of this type is "Kellek," a word of Syrian origin.

The North American Indians occasionally made canoes covered with moose or caribou skins, but only when birch bark was not conveniently obtainable. No trees grow large in the far north; consequently, large pieces of birch bark are not to be found. The northern Indians, lacking birch bark suitable for canoe covering, tried caribou skins.

The Kutchin tribe, of the Yukon Territory and of Alaska, made an open, caribou-skin-covered canoe almost intermediate between the Eskimo kayak and the birchbark canoe. Similar skin-covered canoes are used by the Tahltan Indians of the northern Mackenzie River basin. Some of these canoes are partly decked over like a primitive kayak. In any case, the use of skin for covering canoes by Indians was either only occasional or, if customary, was confined to a few tribes living far north, close to Eskimo neighbors.

Through clean open tracts of virgin white pine and tangled jungles of swamp alder in unexplored Newfoundland in 1874, Mr. T. G. B. Lloyd, a surveyor, led a party of Micmac Indians. Knowing nothing of the country to be traversed, they started on

Fig. 30. Sealskin float from Chile, the South American kayak.

foot, back-packing all necessary equipment and supplies. Before long they discovered lakes, and Lloyd wished he had brought a canoe. They were living largely upon caribou meat and had the rawhides available. The Indians set about making a canoe framework out of green spruce and balsam boughs, bound together with spruce roots. Over the outside they stretched three shaved skins of caribou, sewn together and sewn to the frame with sinews taken from the back of the caribou. The canoe was seventeen feet long, four feet amidship, and carried easily a load of 600 pounds and four men. With that canoe they traversed the lakes and ran down the rapids on their way out to the coast. When they quit work for the season, the Indians ripped off the caribou skins to be used for making snowshoe strings and moccasins during the coming winter.

In South America there are not many native skin boats, perhaps because there are plenty of trees for dugouts and bark canoes, and abundant reeds for balsas. Moreover, there are not many native animals providing skins stout enough for watercraft covering. Even the *pelota* appears to have come into use only since the introduction of domestic cattle.

On the coasts of northern Chile the most important seagoing craft was made of blown-up sealskins. The shores of Chile are washed by waters of the Antarctic Ocean currents, which are much colder than might be expected considering the warmth of the air. In these cold waters lives an abundance of fish and seals. The natives devised a float composed of two long cylinders of inflated sealskin united by a platform. (See Fig. 30.)

In making these floats, the natives first soaked the rawhides in fresh water until they became soft; then they stitched them together into a sort of great bolster. They laid the edges of skin to be joined side by side, and stuck fish bones or stout cactus spines through the hide near the edges; then they drew the two edges firmly together by winding threads of seal intestine around opposite spines. Leaving an opening in the top of each bag, they filled the bags with dry sand or rushes to keep them distended, and set them in the sun to stiffen into shape. Finally they emptied out the sand, sewed up the opening, inserted a small reed or intestine mouthpiece in the top, coated the whole float with seal fat, and painted it with various coats of clay and oil. The floats thus made were not merely waterproof but as airtight as balloons. In case they became deflated while at sea, the canoeman was able to blow them up again through the mouthpiece provided for that purpose.

Each cylinder was between seven and ten feet long and was made from the skins of four or more male seals. In some cases the floats were fastened together in parallel position, but then the platform was so wide that the canoeman could scarcely paddle on both sides unless he used a tremendously long-handled paddle, and that is very awkward. They usually placed the floats together at one end, and allowed them to spread apart at the other end. Due to the taper of the floats the bow was narrow, thus affording easy reach to the water by a paddler kneeling on the platform. At the same time the wide stern gave a sort of triangular shape to the craft, making it impossible to overturn, a wonderful boat in which to outride a storm. The crew consisted generally of two men. The paddler used a double-bladed paddle and knelt near the front; the other man with a long-handled dip net fished from the stern.

Ordinarily the canoemen were supported well above the cold sea water, and remained dry and warm on the sealskin float, whereas they would be wet and cold most of the time on either a balsa float or on the sailing raft, jangada, used by other natives in the same district. This was an important consideration because these natives are subject to pulmonary diseases.

As the canoeman paddled his triangular float, the bow would swerve from side to side with alternate sweeps of the paddle. If ever a double-bladed paddle was needed to keep a craft on a true course, it was needed for that sealskin float. For all that, it was the only skin boat used in all South America so far as I know, and was certainly not copied from anybody else. Unfortunately, these craft are now obsolete. S. W. Lothrop says that he saw in 1929 the rotten remains of what was probably the last one in existence, outside of museums, and at that time it was said there were only two men alive who knew how to make them.

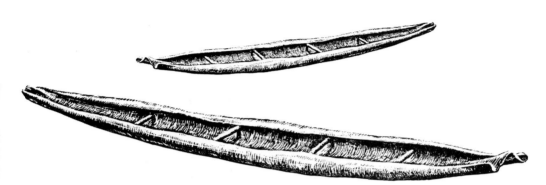

Fig. 31. Jatoba bark canoes, central Brazil.

Canoes Made of Bark 9

An Australian sea hunter stands in the bow of a canoe poised for action. He holds in his right hand a heavy harpoon with bark rope line attached. Searching the waters ahead he spies the quarry — a great aquatic mammal, the dugong — submerged to the nostrils. With his left hand he motions to his crew behind him and suddenly springs out of the canoe and drives the harpoon with his full weight into the dugong. Then follows the full fury of pursuit and the final dispatch of the dugong by the other hunters; the harpooner scrambles back into the canoe however and whenever he has the opportunity.

Some of the canoes used by the Australian sea hunters are made of bark. Bark, and bark alone, has been used for making canoes in many parts of the world. The best known is the birchbark canoe; however, there are many canoes which are made of bark other than that taken from the birch tree.

The simplest sort of canoe is made by tying up the ends of a cylinder of bark, with a few sticks placed as thwarts to hold open the upper side. This is called a woodskin, under whatever variation it may be found in various parts of the world. It has no inner framework, either for strength or for control of its shape. It is much cruder than a bark-built canoe, within which an inner framework of rigid construction is built to conform with the shape of the bark covering. Occasionally a rigid framework is sheathed over with bark, much as the Eskimos first build the kayak frame and then cover it with skin. This is an uncommon system for making bark canoes, requiring a completely different procedure, and produces a so-called bark-covered canoe.

Woodskins are made from the bark of any tree large enough and tall enough to provide a cylinder of bark that may be stripped off in one piece. Obviously only bark that is pliable and tough, preferably with longitudinal grain and a fibrous texture, is good for the purpose. In Australia, woodskins used to be quite common. They were many bark canoes of this type in the southeastern part of Australia and southern Queensland. Several varieties of the eucalyptus tree grow large enough, and sufficiently clear of lower branches, to provide long stretches of suitable bark; it has a very definitely longitudinal grain, and one variety has a tough inner fibrous bark.

Another tree was widely used: the Dibil palm — locally called the stringy-bark tree. The bark from this tree was let down carefully upon the ground, inner side down. The loosened outer bark was cleaned on down to the fibrous inner bark. It was steamed over a fire until pliable, turned inside out, and the ends were tied together with lashings of the fibrous bark. The ends were steamed, the inner bark was stripped off down to the fibrous layer. The natives folded the ends fanwise and bit the folds to make them hold. They then tied up the folded ends, put a few cross sticks to hold the gunwale edges apart, and completed their canoe. Canoes made of bark turned inside out tend to curve back to their natural shape. The Australians prevented this by putting in cross ties of bark rope as well as cross sticks to preserve the shape.

Bark from the mountain ash and the red gum were not turned inside out, and consequently canoes made from such bark were more round bottomed and less likely to spread out of shape.

Even so, they were apt to become misshapen and were generally short lived. Woodskins made of a single piece of bark vary in size from those barely five to twenty feet long. Only by spreading the bark open so as to make the part amidship flat could any approach to steadiness be obtained. A canoe fifteen feet long, three feet wide, and only eight inches deep probably was fairly steady in the water, but must have been easily swamped when there were any waves.

Other bark canoes were used rather generally by the tribes of northern Australia and Queensland, around the Gulf of Carpentaria, and even on the sea. The Anula tribe, among others, used to make well-shaped seagoing canoes out of bark sewed together. Ordinarily they made either the bottom or one side out of a single piece of bark and sewed other pieces on to complete the canoe. These canoes were usually provided with a slight sapling sewed along the upper edge of the bark for a gunwale, with cross sticks for thwarts, and rope ties about the thwarts to prevent too much spreading. To add strength, the natives placed at each thwart two crossed sticks slanting across with one end at the thwart end and the other about halfway down on the opposite side. They also put an extra layer of stiff bark in the bottom of the canoe. These canoes were pointed at each end, were made with considerable sheer, and were fairly seaworthy. The Australian natives used them extensively for fishing, hunting dugong and sea turtles, and for making long journeys.

With the sewed-bark canoes they used five-foot paddles with oval blades; they also used poles when they could; and in the more simple, single piece woodskins they often used short paddles, one in each hand, or mere rounded scoops of bark without handles. Often they used the sewed-bark canoes for fishing with spears. The men used their fish spears, ten feet long, as poles or paddles for the canoes. When using a spear handle or a pole, the natives usually stood in the canoe, but to use a paddle they often sat on their haunches with the right leg under and the left knee drawn up to the shoulder.

Usually the Australians made a hearth of earth and stones in their canoes, on which they kept a fire going. Often the whole family went together on the fishing expedition, cooking and eating the fish as soon as they were caught, with the women and youngsters crowded around the fire. Many women showed the marks of accidental burnings about the small of the back as a result of these fishing trips.

Perhaps the best woodskins of the world are found in the forest lands of the Amazon River valley. In 1887-1888 Karl von den Steinen, a German professor, found them in use around the headwaters of the Xingu River, which is near the border between Bolivia, Paraguay, and Brazil; and E. F. im Thurn in 1882 found them in use among the Indians of British Guiana near the northern borders of Brazil and Venezuela. In general, the woodskins are used throughout nearly all of this vast interior tropical forest land.

There are two trees particularly well adapted for making woodskins. They are the jatoba, also called the varnish or locust tree, which has a characteristic balsamic resin, and the purpleheart tree. Both of these grow large with a tall, straight, clear bole, free of branches for twenty or thirty feet. Von den Steinen says that the Indians usually cut from the jatoba tree a piece of bark of suitable shape, as large as they can manage to strip off without cracking it too badly. They cut through the bark, and pry it off with wedges, using a fire to heat the bark so that it will curl backwards more readily. The jatoba is an elmlike tree with bark about three-quarters of an inch thick. When the bark is stripped from the tree, the Indians have a long half-cylinder of bark in the natural shape of the tree trunk. The very day they strip off the bark, they fold up and lash the ends of the bark into shape, put in cross sticks for thwarts, and let it set overnight. The next day they put it in the water and use it. (See Fig. 31.)

Ordinarily they fold up the stern end of the bark cylinder with a single reversed band, or they wrinkle the end into two or three longitudinal wrinkles. They do not fold or wrinkle the bow; they merely bend it upward. They place the main cargo well aft in order to keep the bow out of the water. This jatoba woodskin is none too good a craft; for one reason, the bark cracks easily and must be plugged with clay or resin. However, such canoes are extensively used and some of them are twenty-five feet long, two and one-half feet wide, and about fourteen inches deep.

The purpleheart tree is used more often to make a rather superior type of woodskin with the ends cut to shape and sewed or bound up. The wood of the purpleheart is so strong that it is used for building the ice-resisting bows of whaling ships, and the bark is so durable that it will serve as a woodskin for months longer than any other South American bark. The purpleheart has a tough, limber inner bark easily separated from the stiff outer bark. Having described how the natives strip off the bark, E. F. im Thurn says: "From each of the two long sides of this, between 2 and 3 feet from each end, a wedge-shaped piece, the base of which corresponds with the outer edge of the bark, is cut out. The two ends of the whole strip of bark . . . are raised until the wedge-shaped slits meet; and these edges are sewn together with bark rope." They cut these triangular gores out of the stiff outer bark only, allowing the inner bark to fold up unbroken on the inside when the outer edges are sewed together.

The Amazon Indians use long-bladed paddles, with a crosspiece handle; they use a very fast stroke, swinging their paddles high in the air as they flash them from one side to the other. For paddles they go to the "paddle-wood" tree. This tree grows with natural buttresses about the base of the trunk that stick out like planks standing on end. They hack off one of the buttresses and whittle a handle out of one end of the natural board thus provided.

Another variation of bark canoes used in the Amazon River basin is called the *yamamadi*. (See Fig. 32.) This canoe is made from a whole cylinder of jatoba bark, and consequently is deeper than the ordinary woodskin. The ends are not folded but are sewed together along the edges, and thereby brought to a sharp cutwater.

Fig. 32. Yamamadi, a bark canoe, central Brazil.

Still another type is made by excavation of the inner part of the tree trunk through a longitudinal slit down the length of the trunk, leaving both ends plugged with the unaltered woody interior and the middle part spread open with thwarts. Such a craft can be made only from a tree with soft pithy interior, like the palm trees. A common palm in the American tropics is the paxiuba. The Indians cut down this palm and, splitting open the outer sheathing, cut out the inside. This leaves a vessel wide amidship and slender at the ends. Spreading the central part still wider by the use of thwarts, the natives make the ends turn upward, and thereby contrive an open canoe that has pointed and upturned ends. In order to prevent the split from going too far they bind the ends with fiber lashings. A few other canoes made from palms are dugouts, such as the East Indian *dunga*, but the paxiuba holds the distinction of being a perfect transition between a bark canoe and a dugout.

Of these four types the paxiuba is found mostly on the Paru River, Brazil, but by no means is it found everywhere that the paxiuba palm tree grows. The jatoba bark canoe is prevalent in the southern part of Brazil, in the upper waters of the Xingu River. The woodskin proper, with the sewed ends, seems to be restricted to British Guiana, Venezuela, and the extreme north of Brazil. The *yamamadi* is most prevalent in the west of Brazil.

Woodskins are also found in Africa. Photographs taken near Cangamba in eastern Angola, Africa, show a fisherman of the Vachokwe tribe using a woodskin canoe. The woodskin was made of a single piece of bark with longitudinal grain, the ends being sewed closed with raffia. The craft was open for its full length; it had neither interior frame nor thwarts. It appeared to be an extemporaneous affair made with no expectation of prolonged use.

Many fishermen use this craft to paddle out to midstream on the Kwando River. They fish with a little butterfly net, held open by a piece of stick about fifteen inches long. The canoeman kneels on the bottom and paddles with a very crude paddle made out of a blade which is held in a slot cut in the end of the handle. Since there is no seat, and no thwarts to hold the shape of the canoe, the top of the back curls over at the edge, where a gunwale might be expected.

From the Pungwe River, about eight hundred miles east of Cangamba, in the northeastern part of southern Rhodesia near Portuguese East Africa, the use of the single bark canoe has been reported. This canoe, like the Cangamba woodskin, was also made of a single, continuous length of bark, with the ends sewed into shape without internal framework. There was, however, the beginning of a framework. There were poles for thwarts and crossed diagonals at the gunwale level. This canoe goes beyond the Brazilian woodskin of purpleheart bark by having diagonals for bracings at gunwale level. A few other bark canoes of similar type have been reported from the neighboring Zambezi basin.

On the shores of the Indian Ocean and at the port of Mozambique there used to be seen, as late as 1910, square-ended, nearly flat-bottomed bark boats made of a single piece of bark. The bark remained right side out, a practice apparently general for African bark canoes, and tended to curl up at the edges. The sides were held apart by four sets of three poles in a group; every pole was held in place by being stuck through the bark. These boats were punt shaped and were propelled with poles.

The bark-covered canoe of Mozambique is characterized by having a framework upon which the bark covering is attached. The function of the bark is merely that of a waterproof covering, whereas the shape, rigidity, and mechanical coherence depend upon the framework. The Mozambique bark canoe is built with a framework which consists of a large number of closely set transverse frames made of bamboo poles bent into a wide U-shape. (See Fig. 33.) These are held in position and stiffened by (a) two pairs of stout bamboo poles running longitudinally on the floor, each pair lashed on at about nine inches on either side of the median line; (b) another pair on each side, forming a rude but efficient gunwale; and (c) a single and lighter bamboo made fast to the ribs on each side, four to five inches below the gunwale pair. Five stout plank thwarts, each about nine inches wide, strengthen the framework transversely. A hole in the middle thwart, together with

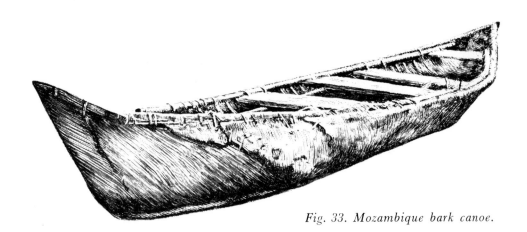
Fig. 33. Mozambique bark canoe.

a short wooden bar on the floor below, indicates that the canoe can be used under sail.

The bark covering is in long, thick sheets sewed together with palm-fiber cord, or coir yarn, coconut fiber. The sides are made up of several lengths of bark, the end of one piece overlapping its neighbor wherever there is need to join two together. The bark skin is sewed to the bamboo gunwales by double stitches at short intervals, with a single connecting cord running longitudinally on the outside. A specimen of this type of bark canoe from Mozambique is in the Museum of Anthropology in Berlin.

The Yahgan Indians of the Straits of Magellan and of Tierra del Fuego, a terribly depressed and underprivileged tribe, have achieved a really notable bark canoe. None of these bark-built canoes are still being used, but there are pictures available and models preserved in museums. The canoe was built out of three, five, or more cigar-shaped pieces of bark sewed together with whalebone, sealskin, or strips of twigs. The cracks between the joined edges were caulked with rushes and resin. The bark was cut from the evergreen beech.

This tree has a bark about one inch thick and is exceedingly stiff. But the Yahgan Indians used to cut or chip off the outer rough bark and scrape away the soft inner bark, leaving a strong tough sheet about one-quarter inch thick, which they put on the canoes. The bark strips had to be cut very carefully to patterns; otherwise the canoe could never have a suitable shape. The sides of the canoe were made high and slightly flaring in order to turn the rough seas. The bark was strengthened with a split sapling for a bulwark along each side, to which were attached, by whalebone lashings, closely spaced bow-shaped saplings for the ribs, which of course were requisite to maintain the shape. Across the top were lashed a number of crosspieces of *lena dura* wood as thwarts, generally six or more, which added greatly to the rigidity of the whole. The Yahgan Indians made a lashing between the pointed ends and the nearest thwart, which served to hold the tip of each end to the framework and to hold the bark pieces together at the end, where at least three sections join. There appear to have been no stem or stern pieces to give the pointed ends any protection beyond that provided by the strength of the bark.

The bark had to be stripped from the trees when the sap was running (October or November in Tierra del Fuego), which caused the bark to become loose. A canoe ordinarily lasted less than six months.

The canoe was round bottomed, and had a marked sheer; that is, it curved upward all along to the very end. The ends stood definitely above the level of the center of the canoe, the whole craft conforming somewhat to a half-moon shape. This made necessary the use of paddles about seven feet long, to reach from the high gunwale down to the trough between waves from either end of the canoe.

When a landing was made, the canoe was brought to the beach bow on, and the man of the family went ashore first. His duty was to make a fire while the rest of the family unloaded the canoe. Then the wife threw ashore her clothes and paddled the canoe to the nearest bed of seaweed. She wrapped a strong strand of seaweed to the handle of the paddle, thrust the paddle handle under a thwart, and swam ashore. The blazing fire warmed her and she was then ready to dress and prepare food for the family.

The beech bark is rather brittle, and the beaches of the Straits of Magellan and of Tierra del Fuego are covered with coarse gravel that is sure to smash a canoe in case the waves dash it on the shore. Thus, the precious canoe was safely moored at sea.

The men watched the weather signs and in calm spells ventured far out to sea to follow the porpoises and to hunt seals. Only when stormy weather persisted did the Yahgans seek shelter on land, for they were more likely to starve on land than at sea. Some writers state that the Yahgans in their bark canoes used to hunt and kill whales. I think it quite wonderful enough that they could kill and recover the carcasses of porpoises.

The Yahgans were people without social ambitions and really followed the simple life. Clothing was used solely for protection against rain and cold. They usually rubbed themselves with oil or grease to shed the rain and to help conserve the natural warmth of their bodies. Everywhere they went they carried fire. On land they always kept a fire burning, and even in their canoes they made an earthen fireplace to keep themselves warm. This practice of keeping fire burning gave their country its first European name — Tierra del Fuego, "Land of Fire."

The advantages of the beech-bark canoe to primitive people like the Yahgans were very great. For the most part they had no steel tools, and ordinarily used shell or obsidian knives. With such tools they could fell large trees only by great labor, but they could easily strip off bark and cut and split the saplings required for the framework. The use of whalebone for bindings is reported by almost all observers; the Yahgans either got the whalebone from stranded carcasses or else they actually did hunt and kill whales.

The nearest comparison with similar canoes comes from the woodland Indians of North America, especially the Iroquois, who made canoes out of elm bark, buttonwood, and basswood. Such canoes were made by the Iroquois of central New York State and southward, where these trees were more plentiful than the paper birch. The Iroquois canoe is everywhere stated to have been heavy and clumsy, inconvenient for portaging, and short lived. It was so poor a craft in comparison with birchbark canoes that the Iroquois used to trade eagerly with the Algonquians for birchbark canoes.

Even in the birch-bark country, however, elm bark was occasionally used for canoemaking, especially among the Menominee living on the west side of Green Bay in Wisconsin. The

Menominee were not active travelers and were satisfied with a poorer canoe.

In making an elm-bark canoe, it was necessary to strip off the bark in one piece. The splendid growth of the American elm, with its noble height and clear bole, was an excellent provider of large pieces of bark. The bark had to be turned inside out and made to curve backward in order to face the smooth inner side next to the water. That is said to have been a difficult trick. No wonder the elm-bark canoes were restricted in use; they were difficult to make and inferior in performance even when carefully made.

On the western coast of North America the birchbark canoe was replaced in the north by the skin boat, and in the south by the dugout, but in the interior, was another very important and peculiar craft, the Kootenai pine-bark canoe.

The Kootenai canoe is made of the bark of pine, spruce, or cedar. For the most part they were made from the bark of the western white pine. Their shape is peculiar in that they are longer at the keel line than at the level of the gunwale. At each end the canoe comes to a point which runs out like the ram of ancient ships or like the bow of the Union ironclad "Monitor." The origin of such a shape is in doubt. It has no imitations, and there are no other original applications of a similar design except among the birchbark canoe-makers of the Amur River in Siberia and Manchuria.

A perfect cylinder of bark turned inside out will tend to spread back to its natural shape unless prevented from doing so. The Indians apparently first cut the bark off in strips with parallel ends. Then they sewed the ends up in order to make the pointed bow and stern. The spread of the bark naturally tends to draw together the tops of the sewed ends. Evidently the distance between the ends of the outcurved gunwales is less than the length of the straight keel line. Consequently, the tops of bow and stern take a natural reversed position if the bark is cut off square at the end. There is also considerable strain on the sewing. By making a longer sewed edge, the strain at any one point is reduced as the total bearing surface is increased. To relieve the strain and to strengthen the bark sides, the Indians placed a double gunwale strip along the top on each side, with one strip inside and one outside. They also placed an additional gunwale strip to cover the edge of the bark between the inner and outer gunwales and the upper edge of the gunwales. Thus there was built around the canoe a very strong composite gunwale to control its shape. The end pieces sloping up and back from the line of the keel merely followed the natural shaping of the bent bark. Custom probably was responsible for exaggerating this slope, beyond the natural adjustment of the bark, into the peculiar pointed bow and stern which have persisted.

The Indians who live around the upper waters of the Columbia River near the boundary of the United States and Canada have never had a reputation for great inventive ability or for special intelligence. It is much more likely that they have merely drifted into the adoption of this form of canoe, than that they have invented it. In the first place, the downward pointed bow is no advantage for any kind of canoe work. In paddling upstream there is always a tendency for the bow of the canoe to sink; a downward sloping bow makes the tendency to submerge almost suicidal. In going down fast water with a Kootenai canoe it is certain that the standing waves at the foot of a rapid would come pouring in over the stern as the canoe shot past. The long straight keel line would make a canoe difficult to turn, and unless it were to be used for sailing, which is quite out of the question, the long keel line is a distinct hindrance. It seems clear that only an isolated group of people not given to much travel, not particularly inventive, and under the heavy pressure of custom and tradition would tolerate such a canoe design.

The Kootenai canoe is open only three-fifths of its length, and when built for one man only weighs about fifty pounds. Canoes about twenty-four feet long, four feet in the beam, and two and one-half feet deep were typical of those used among the

Coeur d'Alene Indians in 1862. A Kootenai canoe of this size held two persons and their luggage. It should, for it is a large canoe.

The Kootenai canoes are reported to have been made by the squaws. Their practice was to strip the bark from the pine in the spring when it is tough and pliable enough to be taken off unbroken and limber enough to stand bending wrong way out. Two squaws took four or five days to make a canoe, including the time required to get the bark off the tree. No gentleman would blame these canoes exclusively on the women. Certainly splitting the cedar poles for gunwales, splitting and whittling down the cedar and maple strips for ribs, false ribs, and thwarts is the work which falls naturally to an Indian man. The men of the Salish and Kootenai tribes cannot shirk all responsibility for these curious canoes which are well adapted neither to the waves of Lake Kootenai nor to the rapids of the upper Columbia River.

10 Birchbark Canoes

Birchbark canoes are found in general use not only among the Indians of the north and eastern woodlands of North America but also among entirely unrelated tribes of Mongolia and Siberia. Furthermore, birchbark canoes have served the purposes of far-flung travel for all the early white explorers and missionaries in northeastern North America.

In 1673 Père Marquette left Michilimackinac with only one companion and two fifteen-foot canoes. They traveled south down the Mississippi as far as about the 33d Parallel, before they turned back to their northern base. Marquette died on the way back, near the present city of Chicago. Cavelier de La Salle traveled in four birchbark canoes, which must have been large ones, with eighteen men all the way from Canada to the Gulf of Mexico in 1681-1682, and he returned in the same canoes.

When Mackenzie started out from Fort Chipewyan to the Pacific, he first built a huge birchbark canoe. It was twenty-five feet long within, exclusive of the curves of stem and stern; twenty-six inches in depth; and four feet, nine inches in beam. In spite of this great size it was so light that two men could carry it on their shoulders for three or four miles without stopping for rest. In this craft the party stowed their provisions, trading presents, arms, ammunition, and baggage to the weight of three thousand pounds; the canoe also held Mackenzie, seven white companions, and two Indians.

Birch bark can be peeled easily and in one piece from the trees in North America in June and July. The grain or the direction of splitting of most species of trees runs up and down the tree, but in the birch, the bark tends to split around the trunk. Furthermore it separates easily into sheets, as thin as paper. This characteristic of birch-bark sheeting makes it possible to use layers between one-sixteenth and one-eighth of an inch thick, as compared with the thickness of evergreen beech bark, one-quarter inch, of the Yahgan canoes, and the thickness of the Brazilian woodskin, one-half or three-fourths inch. While the thinness of birch bark makes it possible to save weight in the construction of the canoe, the tendency of the bark to split at right angles to its natural direction of growth is greater and is a serious weakness. In the sewing together of two pieces of birch bark, for instance, the lines of holes must never be placed along the direction of splitting; otherwise the bark will tear along the line of sewing. Holes for sewing must be staggered or must run in a line oblique to the direction of splitting. Among the Saulteaux or Ojibway of the Nipigon region, I have found cedar twigs running along the line of sewing under the stitches of cedar roots on both sides of the birch bark. This arrangement prevents the pull of the stitches coming so severely on the edges of the holes, distributing the pressures all along the surface of the bark.

The Indians use birch bark from morn till night. They use it as kindling, for it will burn even when wet on the outside. They shelter their fires from wind or rain with sheets of bark. They make cups by folding a piece of bark into a funnel shape and fastening the edges together with a split stick. The stick serves as a paper clip and handle. A roll of birch bark serves as a quiver or a feather case. A larger roll holds the papoose. The bottom of this roll is covered with dry, absorbent sphagnum — a moss which serves the baby as diapers.

I remember once hearing piercing rhythmic noises coming from a Saulteaux wigwam. Peering under the wigwam covering I saw a young girl rocking a cradle. The cradle was made of a strip of bark slung from a pole resting on two forked sticks driven into the ground near the fire. At first I thought the noises came from the baby, but I soon realized that it was only the dry bearing of stick on crotch that made the weird and agonizing squeaks.

The wigwam itself was covered partly with a canvas tarpaulin and partly with pieces of birch bark. Pieces of bark, about four feet square, were bound at two opposite edges between split saplings. The split saplings serve not only to hold the bark in position on the windward side of the wigwam, but also to prevent the very delicate bark from splitting into narrow and useless shreds.

Precautions of this sort must be followed consistently throughout the making of a birchbark canoe. For instance, if birchbark canoes are not put under cover during the winter, the bark tends to separate into sheets. Water within the layers of the bark makes the canoe considerably heavier than when the canoe is new and tight. I once carried an old birchbark canoe over a portage one and three-quarters miles in length, and water was still dripping out of the bark when I set it down at the far end of the portage more than thirty minutes after taking it out of the water.

Another peculiarity of birch bark is its extremely inflammable nature. Although it may be advisable to set fires around some trees in order to hasten the uncurling of the bark, it is dangerous in the case of birch trees. In South America the Indians build a fire near the jatoba tree to make the bark unbend from its natural shape around the tree, but fire causes peeled birch bark to curl more tightly. Furthermore, most birch trees are shaggy, with paper-thin shreds of dry outer bark that blaze up at the touch of fire. The fierce burning of birch bark is due to its high content of pitch; it gives off a strong aroma like that of burning turpentine.

The fraility of the birchbark canoe is often much exaggerated. One ethnologist wrote: "They are very fragile, and every day some hole has to be stopped with gum." I am happy to testify to the contrary. It is true that the Malecite Indians occasionally lash a layer of spruce bark or thin strips of cedar on the outside of the birchbark canoe to protect it from being smashed against the rocks in streams with many rapids. However, this is done only where the rocks are especially jagged and where any other boat would need additional protection. In general, the birchbark canoe is a sturdy, handy, and very reliable craft; with proper care it may give good service for several seasons.

The Indians usually make their canoes during their annual trip to the trading post. After spring has come, they cache their winter belongings; pile the children, dogs, furs, and camp equipment into the canoe; and head for the Hudson Bay Post. The

Hudson's Bay Company has tremendously affected the life of the Indians. At the Hudson Bay Post they camp as close to the trading stores as the factor will permit. They turn up year after year, each family from its own "country." A good Indian knows his own country and all it supports, and takes only those pelts which natural increase can replace. The annual trip to take in the furs and buy the winter's supplies is a happy picnic for the whole family.

Making the new canoe is a social event. The man leaves his camp and goes off in his old canoe to collect clear birch bark for covering, good straight maple poles for thwarts, choice cedar for gunwales, ribs, and sheathing. While he is gone, his wife and daughters collect root fibers for the sewing and balsam fir gum for the caulking.

After materials have been assembled, the Indian marks the outline of the canoe with stakes set nearly vertically in the ground and arranged around an old canoe used as a model. A single piece of bark is used as the bottom of the canoe, and upper pieces are sewed to that for the sides and end top pieces. The whole bark shell is thus placed within the stakes, and the gunwales and ribs are inserted within their confines. The bark is sewed first to the gunwales and last at the ends, and the ribs and inner sheathing of cedar are put into place. Four men work at each end of the canoe. They act as a team and work toward the center of the craft when cutting and shaping the bark and when putting in the ribs and inner sheathing. Meanwhile the women sew the bark, following the gores cut out by the men to make the bark fit the canoe shape. A team of sixteen Menominee Indians, eight men and eight women, have been known to cut, fit, and sew the bark to the framework within a single day. They apply the bark while watersoaked, when it remains pliant enough to take any shape into which it is bent. After a day in the summer sunshine, however, the bark dries out, becomes stiff, and sets into shape. When thoroughly dry, it is too brittle to be formed into any new shape.

Among many Indian tribes the women used to boil split spruce-root fibers in fish broth before they were used for sewing the bark. The fish provided a weak glue to help hold the fibers in place after sewing, and the heat rendered them more pliable. The women also poured boiling water over the parts of the bark which required the most sharp folding; this induced great pliancy and toughness under the strain of bending. The rosin or gum used for caulking was boiled with powdered charcoal to thicken and to color it. The charcoal probably absorbs certain volatile constituents of the crude pine or spruce gum and renders the product less liable to remain sticky after application to the canoe. In order to prevent the gum becoming so brittle by boiling that it chips off too readily, the Indians add grease and sometimes red ocher. Among all Indians the overlaps of bark are so arranged that the exposed edges face toward the stern so as not to reduce the speed of the finished canoe.

While at work the Indians laugh and chatter like magpies. The idea that Indians are stolid, untalkative, and rather unsocial comes from white people's misinterpretation of their good manners. To the Indian good manners require great reserve and very little speech at the first meeting; at later meetings reserve may be dropped, but white people seldom take time for more than one visit.

Considering that birchbark canoes have been used by Indians from the Arctic Circle to the Middle West, and from the coast ranges to the sea routes of southern Labrador, it is clear that there must be different types of birchbark canoes made and designed locally to meet special requirements.

In the far north the Indians are in contact, generally violent contact, with the Eskimos, and both use canoes. The Eskimos use a canoe having a rigid frame independent of the skin covering, but the Indian builds his canoe frame to conform to the natural shape taken by the much stiffer bark covering. The Eskimos must have a deck covering to keep out the splash of waves on the open sea, but the Indian using his canoe on inland waters usually needs no decking. He meets that problem ordinarily by building an upturned bow and stern.

There were, however, certain Indian tribes in the east who used birchbark canoes on the sea. The Micmacs of New Brunswick used a humpbacked canoe on the Restigouche River and at sea, and the extinct Beothucks in Newfoundland once used a high gunwale amidship in order to keep out the waves.

The Indians on the Yukon River and its tributaries in northern Alaska, the Tinneh tribe, used a birchbark canoe having raking ends, instead of the usual recurved ends, and with about five feet of one end decked over. (See Fig. 34.) The decking is obviously an adaptation of the Eskimo deck covering. The ends also are like those of the kayak, in some cases even extending into a sort of post above the level of the gunwale ends. The birch bark, instead of being sewed to the gunwales, is lashed at intervals with root fibers or animal sinews.

The framework differs from that of the ordinary birchbark canoe in having the ribs spaced much more widely. Ordinarily the ribs are about a hand's breadth wide and are spaced at the same distance apart. With the Athapascan (Tinneh) Indians the ribs are much narrower, widely spaced, and are provided with narrow, widely spaced siding and flooring strips much more like the longitudinal frame of the Eskimo kayak than like the almost solid sheathing of the eastern Indian bark canoe. That this is no accidental or special case is borne out by the fact that other Athapascan Indians (Dog-ribs) living on the Mackenzie River use a similar model, except for the fact that ordinarily both ends of the canoe are decked over for a distance of between two and three feet.

The Indians invented two main types of birchbark canoes with all gradations between. One is a canoe with a pointed but not recurved end, the line of bow and stern in profile rising gradually to an upper point which is at the extreme length of the canoe. In the other type the extreme length of the canoe is at a point between the ends of the gunwales and the end of the keel, with a

Fig. 34. *Athapascan canoe, Yukon River.*

marked concavity in outline between keel and top of the canoe. The first is illustrated by the type developed in New Brunswick by the Malecite Indians, and the second by the Saulteaux of northern Ontario. The Malecite add a short covering at the end of the gunwales, like a very short decking, and a bark flap below the gunwale on the outside of the bark covering at each end. This flap serves to deflect the upward splash of water from going into the canoe. The use of this flap and inner decking becomes less marked among the tribes farther west toward the interior. The Montagnais of northern Quebec, who make a canoe with recurved bow line, use no outer flap, and neither do most others who have a bow with a high recurved end. They do not need it; the high end serves the same purpose.

Occasionally an exaggerated form of the recurved end is seen, resulting in a continuation of the bow edge in a horizontal direction. Certain canoes of the Chipewyan have this type well developed, but probably the Slave Indians from the Hay River which flows into Great Slave Lake have built the most exaggerated upturned ends on their canoes. Their canoes are shallow, but have very high ends, with a horizontal edge and a small decking which makes the ends of the canoe boxed off from the main part. Because no ribs can be inserted in these exaggerated ends, it is customary to stuff the end with shavings or moss in order to keep its shape and in many cases to close the end with a thin slab of cedar. The highly upturned end is most picturesque and was exaggerated in the trading canoes used by the old *courriers de bois*. Tassels were sometimes tied to the point and figures or emblems painted on the side of the bow.

In modern canoes an outer gunwale is put on as a protection against wear by the paddles. It adds stiffness to the birchbark canoe and certainly protects the sewed attachment of the bark where it is most vulnerable, but it also increases weight somewhat. The Slave and Athapascan Indians attach the bark between two gunwale strips, but use a binding around all three units at intervals — probably the Eskimo influence. Most of the Indians, from Great Slave Lake to the Huron and Iroquois country to the south and even to the eastern sea coast, used an inside gunwale to which, in true Indian fashion, the bark was sewed continuously along its length.

Birchbark canoes designed for use at sea have certain notable differences. It is desirable that they should be held to a course in spite of waves and winds and that there should be protection again waves. The Beothucks of Newfoundland accomplished this by means of a keel-like bottom ridge, in contrast to the almost universal flat bottom of the river canoes. Water is kept out of a canoe when it heads into a wave by the high bow, but when a canoe rides over a steep wave amidship, the crest of the wave overflows the gunwale and water pours in. That is why the Micmacs and Beothucks made their canoes with a raised hump amidship. The humped gunwale kept out the waves amidship. In order to avoid the slopping in of water when the canoe rolls sideways in the waves they made the sides of the canoe slope upward and inward, a so-called "tumble-home" shape. This shape is common among the canoes of the seagoing Micmacs and the Passamaquoddy tribe of Maine, but is practically unknown among the inland Indians.

Another device for travel at sea or on large lakes is an increase in the size of the canoe, including a higher side. The woodland Indians used great canoes holding fifty or more warriors for use on the Great Lakes. The warlike Iroquois, for example, actually paddled across the Huron and Ontario lakes in order to attack their enemies, the Huron Indians. These war-canoes were copied by the French fur traders and by the servants of the Hudson's Bay Company for freighting their commodities, and were further developed into the very high-ended large canoes called the "rabiscaw." In our own times the Indians of Alaska and the Micmacs of maritime Canada make birchbark canoes that have very high gunwales for use on the sea.

The eastern Crees made canoes twenty-five and thirty feet long, with which they traveled along the coasts of Labrador. Their paddles were short and rather heavy, and the paddlers took short, jerky strokes in comparison with those of the neighboring Ojibways. When men paddle over waves in a long canoe they cannot paddle effectively in unison if they try to take long strokes, because some will paddle deep in the wave tops and others will skim the water in the troughs. A long paddle is of no help if the paddler has to use the short, jerky stroke which is most effective on rough or choppy water. Consequently, for travel on sea the Indians changed paddles as well as canoes

Perhaps the most remarkable seafaring birchbark canoe ever developed was that of the Beothucks of Newfoundland. T. G. B. Lloyd, a surveyor, wrote the following account in 1874:

> The principle on which the Red Indian's[4] canoe is constructed is perhaps nowhere else to be met with. It has in a way no bottom at all, the side beginning at the very keel, and from thence running up in a straight line to the edge or gunwale.
>
> A transverse section of it at any part whatever makes an acute angle, only that it is not sharpened to a perfect angular point, but is somewhat rounded to take in the slight rod which serves by way of a keel. This rod is thickest in the middle (being in that part about the size of the handle of a common hatchet), tapering each way, and terminating with the slender curved extremities of the canoe. The form of the keel will, then, it is evident, be the same with the outline of the longitudinal section, which, when represented on paper, is nearly, if not exactly, the half of an ellipse, longitudinally divided. Having thus drawn the keel, whose two ends become also similar stems to the canoe, the side may easily be completed after this manner; perpendicular to the middle of the keel, and at two-thirds the height of its extremities, make a point; between this central and the extreme points, describe each way a catenarian arch, with a free curve, and you will have the form of the side, as well as a section of the canoe, for their difference is so very slight as not to be discernible by the eye, which will be clearly comprehended on recollecting that the side, as I said before, begins at the keel. The coat, or shell, of the canoe is made of the largest and fairest sheets of birch bark that can be procured, its form being nothing more than two sides joined together, where the keel is to be introduced. It is very easily sewn together entire. The sewing is perfectly neat, and performed with spruce roots, split to the proper size. The portion along the gunwale is like our neatest basket-work. The seams are payed over with a sort of gum, which appears to be a preparation of turpentine, oil, and red ochre, which effectually resists all the effects of the water. The sides are kept apart, and their proper distance preserved, by means of a thwart of about the thickness of two fingers, whose ends are looped on the rising

---

[4] Although Lloyd called the Beothucks "Red Indians" because they used so much red ocher in painting their faces and bodies, the term is merely descriptive. Lloyd was apparently unaware that there was also a tribe in Newfoundland quite different from the Beothucks but also called "Red Indians."

points above mentioned in the middle of the gunwale. The extension caused when this thwart is introduced, lessens in some degree the length of the canoe by drawing in still more its curling ends; it also fixes the extreme breadth in the middle, which is requisite in a vessel having similar stems, and intended for advancing with either of them foremost, as occasion may require, and by bulging out their sides gives them a perceptible convexity, much more beautiful than their first form. The gunwales are made with tapering sticks, two on each side, the thick ends of which meet on the rising points of the main thwart, and, being moulded to the shape of the canoe, their smaller ends terminate with those of the keel rod in the extremities of each stem. On the outside of the proper gunwales, with which they exactly correspond, and connected with them by a few thongs, are also false gunwales, fixed there for the purpose of fenders. The inside is lined entirely with sticks, or ribs, two or three inches broad, cut flat and thin, and placed lengthwise, over which again others are crossed, which, being bent in the middle, extend up each side to the gunwale, where they are secured, serving as timbers. A shut thwart near each end, to prevent the canoe from twisting or being bulged more open than proper, makes it complete. It may readily be conceived, from its form and light fabric, that being put into the water, it would lie flat on one side, with the keel and gunwale at the surface, but being ballasted with stones it settles down to a proper depth in the water, and then swims upright, when a covering of sods and moss being laid on the stones, the Indians kneel on them, and manage the canoe with paddles. In fine weather they sometimes set a sail on a very slight mast, fastened to the middle thwart, but this is a practice for which their delicate and unsteady barks are by no means calculated. A canoe about fourteen feet long is about four feet wide in the middle.[5]

Certain points about this canoe are most interesting. In the first place, the high gunwale amidship is certainly a protection against the seas. The Beothucks were accustomed to make occasional trips to Funk Island thirty miles from the main island in order to collect birds' eggs. It seems incredible that anyone would make such a trip in an open canoe, operated only with paddles. Perhaps, as Lloyd says, the Beothucks used a sail fairly consistently. If they did use a sail, the hump shape amidship would be a great protection if they heeled over in a stiff breeze, and the keel-like bottom would help to prevent side drift.

The high sides and lack of obstructing thwarts would make the Beothuck canoe useful as a tent. These people were seacoast dwellers, and their camp often had to be made on the beach or

---

[5] T. G. B. Lloyd, "The Beothucks," *J. Royal Anthropol. Inst. Gr. Brit.,* IV (1874), 26-28.

rocky shores of Newfoundland where a ready-made tent, such as their canoe, would give them shelter anywhere.

A careful reading of the description from Lloyd's writing shows that ribs were bent into shape and placed in this canoe. Its seaworthiness was increased by the wide beam, maintained by the ribs, and kept by the central thwart from becoming too wide. Furthermore, the single strip of keel, presumably made from some springy wood like maple, whittled down to a size of about one and one-half inch in diameter in the middle and tapering to its ends, must have tended to spring the ends of the canoe apart, contrary to the upcurling effect produced by the spreading amidship. The inner gunwale, to which the bark was bound so admirably, was absolutely necessary in order to prevent the bark splitting all around the canoe as it bulged open. No doubt the outside gunwale served a highly important part in strengthening the sides of the canoe even though bound only at separated points along the gunwale line. The gunwales were thickest near the central hump where the strain on the birchbark would be most severe.

North American Indians are not the only birchbark canoe-makers. Many tribes of far eastern Siberia have used birchbark canoes for centuries, having evolved their own designs and methods of handling, quite distinct from those of the North American Indian with whom they could not have had any direct or even remote contact. The white birch grows on the island of Sakhalin, north of Japan, along the Amur River which separates Mongolia from Siberia in the Far East, and in European and Asiatic Russia. Along the Amur River and its tributaries live a group of Asiatic tribes of obscure ethnological affinities; they are known as the Goldi, Orochi, Manyorgs, Tungus, and Birari, and near the mouth of the Amur and on Sakhalin Island, the Giliaks. All these peoples have birchbark canoes of many types, all considerably different from those of the North American Indians.

One outstanding feature is a long keel line, which reminds one of the Kootenai pine-bark canoes, but this is the only obvious similarity. The Asiatic birchbark canoemen prolong the keel into a sort of reversed beak, often spear shaped at the tip. The Asiatic birchbark canoes show much greater variety than the American models. They vary from very long, narrow racers to short, wide, almost tubby canoes. The Goldi, Manyorgs, and Birari use long narrow models. The Birari make canoes up to thirty-five feet long, which are only twenty-six inches wide, no matter how long they are. Four to six canoemen sit one behind the other, and all use double-bladed paddles. These canoes are much too narrow for the paddlers to sit side by side, and under the pressure of double-bladed paddles they must be extremely fast.

Another tribe living on the lower Amur River, the Goldi, also make long narrow canoes. These are finished at each end with a piece of wood carved into a curved continuation of the snout-like pointed keel line. The bark edges are held between an outer and an inner gunwale, and the open ends of the canoe are decked over with birch bark sewed below the gunwale strip to the sides of the canoe. These canoes have three thwarts, and slat-like longitudinal strips held against the bark by broad, thin, neatly finished ribs. The framework is more like that of the Eskimo kayak than like the American birchbark canoe.

The Tungus make a small canoe out of pieces of bark that go completely around the belly of the canoe from gunwale to gunwale. These sections of bark are sewed together, the line of joint going right around the canoe. At each end the bark is folded over and sewed together, forming a snout-shaped ending. A rectangular frame, bounded by two thwarts and two special gunwales, makes a structure of considerable rigidity at the central part of the canoe. Here the canoeman sits, using a single-bladed paddle. To strengthen the bottom of the canoe, and perhaps to prevent punctures, an extra thickness of bark runs lengthwise from end to end and is sewed through the outer bark covering and around wide, flat, and closely set ribs.

Another tubby type of canoe is made by the widespread Yakut tribe. This canoe also has snout-shaped ends, the line from the extreme ends recurving from the bottom upward. The covering is made from sections of bark passed around the canoe from gunwale to gunwale, folded over the ends of the ribs and the gunwales, and sewed into place. The edges of the bark, of course, are also sewed together. The frame consists only of thwarts and gunwales; there are no ribs or sheathing. This must be a very light canoe, easy to make, and wonderful for portaging.

The Yakut canoeman uses a double-bladed paddle that has beautifully made, square-ended, and tapering blades. The canoe is far cruder in workmanship and design than might be expected in view of the rather finished character of the paddles. The Yakut tribe, like many other Siberian tribes that live between Mongolia and the Arctic Ocean, also use wooden canoes, either dugouts or canoes built up from thin boards of poplar wood. These poplar-wood canoes are very similar in design to the birchbark canoe and it may be that the birchbark is really a makeshift copy of the more finished poplar canoe.

The birchbark canoe is used extensively in the Amur, a river system that has over 8,000 miles of navigable waters. It extends southwest into Mongolia, and west, north, and east into Siberia. The Amur birchbark canoes commonly have a snout-like point at each end. The small canoes are covered at one or both ends with birch-bark decking. There is a widespread use of double-ended paddles, at least on the lower waters of the large streams.

The Tungus tribesmen live on the upper waters, in forested country with rapid rivers like those of eastern Canada. They use a single-bladed paddle and often portage their canoes. It would be interesting to know if they carry their canoes with the aid of single-bladed paddles lashed to the thwarts in Indian fashion. They certainly use pieces of birch bark to cover their igloo-shaped huts much as the Ojibways sometimes use birch bark on their wigwams.

Although the Amur tribes are mostly Mongolian in type, they seem to show Aleutian influences in their double-bladed paddles, their partly decked canoes, and their wide spacing of the canoe ribs. It may even be that the exaggerated turn up of the Goldi and Manyorg canoes is a reflection of the recurved bow of the Aleutian skin canoe. In any case it appears to be of no advantage or service to the Amur canoeman. However, in this connection it is wise to withhold judgment, for it is rare indeed that a natively developed, truly indigenous craft is without its special local advantages with respect to each one of its peculiarities.

Unluckily I have no personal knowledge of the birchbark canoes of Siberia. My information comes from other writers, and from models of canoes brought from Asia. Models are not always faithful copies of the full-scale article, and it is not likely that the canoe models are exceptional; consequently, my descriptions of Asiatic birchbark canoes may have appeared sketchy. The Siberian and Mongolian canoemen have apparently developed definite types nearing perfection, much as the North American Indians have evolved the almost perfect river canoe of the Ojibway and seagoing canoe of the Beothuck. Probably the Birari long, narrow canoes, driven with double-bladed paddles, exceed in speed any birchbark canoe ever produced by the North American Indians.

It is truly remarkable that birchbark canoes of such diversity of form, and such versatility of usefulness, should have been developed independently both in Siberia and in North America. Whatever the reasons, certainly the American Indian borrowed almost nothing from the Arctic Eskimo whereas the Siberian canoe was much like that of the Aleutian Eskimo. In both cases, however, they developed independently excellent canoes of many varieties covered with the wonderful bark of the white birch.

11 Sailing Canoes

A simple canoe is not at all suitable as a sailing boat. It is too narrow to be steady, too low to be seaworthy, and too shallow to sail close to the wind. These are serious defects, but once they have been overcome the native makes a sail and uses it. Native seamanship has always kept abreast, and considerably ahead, of native shipbuilding.

The American pilot boat, with heavy ballast beneath a deep keel and narrow hull, illustrates one way of giving great stability to a sailing ship. The racing yacht follows the same principle. An altogether different type is the familiar fishing smack with wide beam, bluff bows, sharp bilge, broad bottom, and roomy deck. The racing yacht and the fishing smack are represented in principle in various primitive sailing craft. Extremes are illustrated by the Johore fast boat, the amidship section of which is like a yacht; by canoes with one or more outriggers in order to increase the beam; or more simply by two canoes fastened together, side by side.

The Yankee clipper ship owed its speed to its streamlined stern and fine lines. Nearly all the contemporary European sailing ships had square or very bluff sterns. The Johore fast boat preceded the clipper ship and had long been built with a wave-form hull, a design finally made known in Europe and used for the later fast-sailing ships. The sailing canoes and related craft of Malaya were so fast in comparison with European sailing ships that until the advent of the steamboat, piracy and smuggling by the natives continued uncontrolled. European gunboats and frigates could not safely follow the shallow-draft outrigger canoes over the reefs and close to shore, nor could they overhaul the swift sailing praus on the open seas. Malayan pirates had long been notorious — not that they were more bloodthirsty, more rapacious, and more warlike than other uncivilized races of the world, but because their fast-sailing craft made it possible for them to carry on their raids relatively safe from pursuit and punishment. Among the swiftest craft is the Johore fast boat, whose speed depends largely upon the fine lines and wave-form of its deeply dug-out hull.

Most European boats have high bulwarks or deep hulls in order to secure seaworthiness. This design in canoes, however, leads to a top-heavy structure that is too unstable for sailing. Canoes by their very nature are long and narrow; a high superstructure on a canoe is impractical. Even a high sheer, fore and aft, becomes a menace to stability when the wind is high. One obvious precaution is to deck over as much of the canoe as possible, thereby keeping out some of the waves. A more elaborate plan is to build up the sides of the canoe, steady it with the addition of outriggers, and construct a platform for cargo and crew at an elevation considerably above the top of the canoe proper.

A sailing craft must be sturdy enough to carry the severe strain of a large sail, and it must provide such resistance of the hull in the water that the boat sails in the direction of its length rather than drifts sideways. Flat-bottomed or shallow boats make serious leeway when sailing close to the wind. This is corrected either by attaching a deep keel, or by providing a leeboard, a temporary keel, which may be let down alongside or through the center of the keel line. Canoes are seldom built with deep keels, an exception being the curious Bombay fishing canoe called a "machva."

The machva is a dugout canoe pointed at each end, with a rather definite sheer fore and aft, and a well-marked wave-form in its hull. This canoe has a short mast and a curious lug sail, almost lateen in shape. Its most remarkable feature, however, is its keel. The keel is deepest at stem and stern, the profile of the keel rising to the bottom of the canoe amidship. When this canoe is sailing close to the wind it has the greatest grip against side drift exerted at the extreme ends. This aids the canoe to sail close to the wind, maintaining a straight course, but results in making the craft extremely slow in responding to the tiller when it is brought up into the wind. This canoe sails splendidly when close hauled, but it cannot tack. When it reaches the end of its course it cannot be turned into the wind, swung about, and brought back on the course; it loses too much headway during its long, slow swing into the wind. Consequently, the canoemen swing away, with the wind astern, at the end of the course, and "wearing about ship" come up on the other course, never losing their momentum at the turn. Among most sailors this method of sailing against the wind is used only with very light winds or under unusual conditions.

This manoeuvre requires the canoemen to drop the sail, unship the mast, and step the mast again with the sail on the other side of the mast. Luckily the sail is hoisted easily and quickly by the main halyard, and with a crew of four or five men the short mast is easily and quickly handled. In case they lack sea room for sailing against the wind, the crew can row the canoe, using round-ended long paddles which they adjust like oars against wooden tholepins.

The Indians of Peru, South America, invented an ingenious means of steering their sailing rafts. They used leeboards, a device never used on sailing canoes. The Peruvian rafts, jangada, were made of balsa wood, a wood so light and pithy that a young boy can carry a large log on his shoulder. Such logs are extremely buoyant and float high in the water. The jangada were made usually of five and sometimes of seven logs; the logs were lashed together with the longest in the middle and two or three shorter ones on each side. Arranged in order of length, the logs made a very rough point at both ends of the raft. The lashing of the logs was such that the logs could move in relation to one another. The sail was supported by a shears mast, two poles spread at the foot and lashed together at the head, each being footed in a different log. The sails in ancient days were made of reeds tied together in parallel position, forming mats or screens. In spite of the great buoyancy of balsa logs, such a raft lacked freeboard so seriously that waves would repeatedly sweep over it from end to end. The loose attachment of log to log was intended to permit at least a slight lift to a side sea, but it could not be effective in keeping the tops of the logs dry except in calm weather.

When the wind was not astern, the men drove great paddle-like leeboards between the logs, in that way reducing the side drift that otherwise was serious. At the same time they were able to control the general direction of sailing. When the leeboard was thrust down forward, naturally the stern fell away, and the raft headed more sharply into the wind. When the leeboard was put down aft, the bow fell away from the wind, and the raft sailed with the wind nearly astern. The natives actually used these leeboards for controlling the course of their sailing rafts, much as European sailors control the course of a ship by the trim and disposition of sails fore and aft. However, European sailors never developed the idea of steering with a rudder except in the stern, but the South American aborigines used a rudder movable along the whole length of the craft.

Indians have been sailing their large reed canoes, balsas, all over Lake Titicaca since long before the white men came. The balsa rides high out of the water and is given still more freeboard amidship by a humpbacked gunwale. In general, the reed canoe is a success where the jangada failed, that is, in regard to freeboard, but it gains beam and draft only at the cost of great clumsiness. The balsa sailors do not use leeboards like the seagoing Indians, but they steer with the aid of paddles. Although the sailing balsa is slow and awkward to handle, the Indian of today finds satisfaction in letting the wind do the work for him, while he sits in the stern, with steering paddle beneath his knee, and takes his ease.

Canoes that must go out to sea in rough water, or that must outride waves once a storm is upon them, need the extra stability that is provided by an outrigger. The outrigger in its simplest form is merely a log or bamboo float rigidly attached to the canoe by booms, and gives the canoe all the advantages and none of the defects of a raft. (See Fig. 35.) Most single outrigger canoes are small and rather poorly constructed, but locally, as in Ceylon, they are large and capable of carrying up to thirty tons of cargo. In the islands east of New Guinea the outrigger canoes are very seaworthy, are used for long trading voyages, and carry twenty or more natives with provisions and personal equipment.

The canoe is usually sailed with the outrigger on the windward side. The wind pressing against the sail makes the canoe cant over to the lee, raising the outrigger float from the water, so that

Fig. 35. Madura fishing canoe.

it serves as a counterweight against capsizing. Whenever the outrigger is on the leeward side, the canoe is in danger. When the canoes keel over before the wind, with the outrigger on the lee, the outrigger is likely to sink so deeply in the water that the booms break and the canoe is wrecked. When heavily loaded, a large canoe like the Ceylon single outrigger would be stable even without the outrigger; consequently, even when the wind blows on the wrong side, the Ceylon craft is generally safe. These canoes usually creep along close to shore on their coastwise traffic and scarcely ever venture beyond the southern Indian Ocean where the prevailing winds are steady and moderate.

  East of New Guinea the Trobriand islanders build up the sides of their dugout with solid boards, carefully caulked to prevent leaking even when the loaded canoe sinks deep into the water. On the decking between the dugout and the outrigger they frequently crowd a large crew of men. The outrigger is merely a buoyant log used mainly as a counterweight against capsizing, and the main burden is carried by the dugout hull. When under full sail, the canoe seems to lift out of the water because the crew crowds over to the outrigger side to help balance the wind load, thus relieving the weight upon the dugout. Before their voyages the natives pronounce charms or rites of magic to insure this lift or lightness of the canoe. There is no mistaking the obvious delight these canoemen take in the birdlike speed of their craft as it skims over the waves.

The hulls of outrigger canoes are dugouts wherever suitable trees are available. Otherwise a hollow hull is built up out of small pieces of plank of assorted shapes and sizes sewed together with fiber in a most bewildering design. In some localities the pieces are sewed with holes drilled through them, but to make superior canoes, individual pieces are carved with interior ribbing through which the lashings pass. Whenever the hulls are thus built up, the canoe emerges as a sailing craft of fine lines and capable of great speed. The weight of outrigger booms, mast, sails, rigging, and crew sinks the hull deep into the water, making an overdecking or at least wash strakes absolutely necessary. At the same time the deeply submerged hull reduces leeway and enables the canoe to sail close to the wind.

From the point of view of a boatman, the small single outrigger canoe is a stronger and handier craft than a double one of the same size, especially when running before the wind. However, it cannot beat to windward except at great risk unless very special and awkward methods are used. The usual device is to have the canoes double ended, with the mast in the center, and with bow and stern interchangeable. As the boatmen change from one course to the next, they swing the sail over to the other side of the canoe, turn about so that the stern now becomes the bow, bring the bow up toward the wind, and away they sail on another tack. Among some of the islands of the Polynesian groups, a large steering paddle is lashed with a long rope to the mast. A man is posted at each end of the canoe, and as the canoe changes its course the first steersman lets loose the steering paddle which floats past the moving canoe to the second steersman, who then operates from what had formerly been the bow.

Shipwreck of single outrigger canoes occurs only too often. In rough water, the outrigger on the wave crest may be lifted so high above the dugout in the trough, by combination of wind load and wave action, that the whole clumsy thing overturns, smashing the booms and swamping the dugout itself. Then the crew has only the outrigger on which to find support. If far from land, the men soon become weakened by exposure and drown, or are torn to pieces by sharks.

Among the Polynesians, physical comfort is reduced to very simple terms. For the most part, the complete belongings of a family including fishing tackle, cooking utensils, and wardrobe could be packed into a small chest. This scantiness of material possessions makes it possible for the South Sea canoeman to make considerable voyages in his single outrigger in spite of its restricted space. The usual diet of many of these people is a combination of fish and coconuts. The natives catch a fish, bite off the head, and devour the body raw. This solves the problem of carrying fuel, food supplies, and cooking utensils aboard the canoe. Polynesian clothing is so simple that great excitement was created among the natives of one island after one of them had seen the wife of the new missionary undressing before she retired for the night; the report was spread abroad that she took the skin off her legs before she went to sleep. None of the natives had ever heard of stockings.

If this had happened in New Guinea the implications might have been interesting, because most of these natives believe that certain very malevolent witches are women who take off their skins and leave them in their sleeping quarters. The witches then fly through the air seeking the lives of men. Canoemen at sea believe they hear the eerie screams of the flying hags whenever the storm winds sing through the rigging, and nothing can save the luckless canoeman from their clutches except the power of his good strong magic.

The "Kula Ring" is responsible for keeping in active use the hundreds of native canoes that are not actually needed for the subsistence of the people. This Ring is a remarkable trading organization. The main objects of the trade are shell arm bands, shell and bead necklaces, and stone axes — of great native value but without much commercial value. None of these objects have any use; they are not permanently retained by any person or any community within the Ring. They are eagerly sought after, highly prized, but faithfully passed on in the Ring in due season.

An expedition of forty or fifty large sailing canoes sets off on its journey after long ceremonials, magical rites, incantations, and traditional spells. The men carry only their sleeping mats, simple provisions, and articles of magic and talismans. At length the fleet approaches the village which they have planned to visit, and the canoemen carefully perform their rites of magic to make themselves "appear beautiful" or pleasing to their hosts and to bring them good luck and success in the approaching trade. Then they advance in ceremonial state, a youth in each canoe blowing on a conch.

The trade consists of a highly conventionalized, formal presentation and acceptance of native valuables which are given to the visitors with no valuable gifts given in return. This ceremony goes on for some days before the canoemen are ready to depart for their own homes.

The gifts received carry an obligation to return gifts. Each tribe has its own variations in routine, and each year the valuables go from tribe to tribe to be held, prized, exhibited, and enjoyed by each group in turn. New shell arm bands involve searching for the necessary shellfish, and that search requires the use of large sailing canoes.

Sails differ from place to place, but certain types are common. A square sail is strengthened with yards at head and foot, or with one long yard at the head by which the sail is attached to the mast. Another rather common type of sail, especially in the islands of Micronesia, between Japan and New Guinea, is triangular. When Magellan discovered the Mariana Islands in 1521, he named them the Ladrones because the natives were such thieves, but he also described them as the islands of the lateen sails. The sail is held in shape by two long yards on the edges of the sail that come together at the point of the sail in the bow of the canoe. The mast holding this sail in place is set with a sharp slant forward, stepped nearer the stern than the bow. The two yards and the mast incline toward one another at an angle of about 45°.

In southern New Guinea the single outrigger canoes are equipped with a small crab's-claw sail, like those used on the large double canoes of the same locality. On the northern islands of New Guinea, the Jappen Islands, one may see a tripod mast instead of the usual single-pole mast used on single outrigger canoes. The tripod mast is common on praus and other native craft about Java and Makassar Strait; it is not common on single outrigger canoes. The tripod mast is made with three poles, at the head of which an oblong sail is made fast. This oblong sail, like the common triangular ones, has yards attached to both the top and bottom edges by which it is strengthened and made fast to the mast.

Certain tribes build out a balancing platform opposite to the outrigger on which the crew may stand to counteract the partial submergence of the outrigger float. This is in order to reduce the chances of smashing the booms whenever a shift of wind forces the outrigger down on the leeside. This is bound to happen occasionally in spite of the most skillful seamanship. Balancing platforms of this type were, until recently, in common use in parts of Micronesia, especially on the Gilbert and Marshall islands. Obviously a sailing canoe with a minor outrigger and a balancing platform approaches the craft with a hull wide enough to need only a balancing platform, or none at all. Here we see the transition from dugout canoe to sailing ship.

All the dangers of sailing are well appreciated by the South Sea islander. He is an expert swimmer and never fears to drown except as a result of foul sorcery or witchcraft. The Trobriand islanders prepare for their voyages by serious magic observances. The captain speaks spells over ginger root which he wraps up carefully to preserve its strength; he speaks a spell into his lime pot and stoppers it tightly; he takes these and certain magic stones aboard his canoe. The ginger spell is for personal protection; the lime pot is used with spells to cast a mist around the crew so that they cannot be found by the flying witches; the stones are thrown into the sea to make the water turbid so that the sharks cannot find the swimming men, and the "gaping depths" of the sea cannot drown them.

Fig. 36. Large outrigger, Bali.

Double Outriggers and Double Canoes | 12

Early in the Christian Era, great outrigger ships were built by the highly cultured and skillful Javanese. The double outrigger gave room on the booms, straddling the single narrow hull, for cabin space for crew, passengers, and cargo. At the same time the superstructure, erected on booms or poles above a single dugout hull, was much cheaper and less laborious than the construction of equal room in a hulled ship. Today double outriggers have spread west to the east coast of Africa, south to northern Australia, east to New Guinea, and north to the Philippines. In spite of the evident advantages of the double outrigger or sailing canoes, the single outrigger persists for river and inshore service in many places where double outrigger canoes are used for sea voyages.

The double outrigger canoe is the watercraft of a group of people basically different from those who use the single outrigger. For the most part they inhabit islands rich in natural resources — a factor contributing to their higher stage of culture. The natives of the Dutch East Indies, although controlled politically by the Dutch government, live much in their own way, eating their native foods and practicing their native arts and customs. Their diet includes flesh, fish, fowl, rice, sago, yams, bananas, breadfruit, coconut, and various aromatic spices for flavoring. Compared with the diet of fresh or fermented breadfruit, raw fish, and coconut products of the Marquesas, that of the natives of Java, Sumatra, and Borneo is rich, plentiful, and various.

In parts of Borneo the natives hunt deer for meat. In New Guinea they make great drives to slaughter cassowaries and wallabies. The cassowary is a bird without wings, with feathers more like hair than quills, and with great speed as a runner in escaping from danger. The wallaby is a variety of kangaroo, living in the grassy uplands of New Guinea. The natives surround a patch of grass and then set fire to it. As the wallabies come bounding out of the burning grass to seek shelter in the forest they are easily killed by the waiting spearmen. Any cassowaries unlucky enough to be caught in the ring of hunters are killed at the same time.

Wild pigs are almost the only animals to be found in the islands of the South Seas, and they have been imported. This general scarcity of meat probably accounts for the practice of cannibalism from the Solomon Islands to the Marquesas. "Long pig" was the picturesque name given to a human victim. In the Marquesas, said to have produced the most beautiful men and women of the South Seas, human flesh was so desirable that when official enemies failed to supply the demand, villagers were slaughtered to feed their neighbors.

Cannibalism was not generally practiced by the Indonesians, chiefly because they were of a superior culture; they were able to provide themselves with plenty of food by agriculture, trading, and fishing. They were a maritime people, well disciplined to the sea, and capable navigators in their own and in distant waters. They excelled as sailors of the double outrigger canoe and invented many variations for its use.

Outriggers are connected to the hull of a canoe by one or more booms, which in turn are attached to the floats by a variety of methods. Each type of outrigger attachment has definite tribal or racial significance and a restricted geographical range. Consequently, the best way of describing the different types is by grouping them according to the type of outrigger attachment.

Perhaps the simplest is that of the boom with direct attachment to the float. This is probably best known to Americans because it is common in the Philippine Islands on the native canoes, vintnas. These are of different sizes up to a large seagoing craft with two masts and four large booms to carry the heavy floats. Characteristically they have a sharply pointed bow, and an elaborately carved and painted double-ended or bifid stern. In the ornamentation of these canoes, the native builders show remarkable skill as sculptors.

Varieties of canoes using the direct attachment of outriggers are found in east Java, where one type around Madura and the island of Menado uses the boom ends bent downward and lashed directly to the floats.

A second type of attachment is made by use of a U- or O-shaped withe, bent so that the curved part is lashed to the outrigger float and the upper part is lashed to the boom. This variety is generally associated with the islands of Ceram and Amboina, near the eastern end of the Dutch East Indies. A variety of the same type of attachment makes use of a Y-shaped connection. The stem of a forked stick is used for attachment to the float, and the arms of the fork for attachment to the boom. These canoes are widely used locally. Even small children may be seen paddling about near shore in these steady little canoes, using paddles of suitable size. Men, women, and children seem equally at home on the sea.

A third type of connection consists of vertical stanchions driven into the float and through the end of the boom. This attachment is found commonly in use at Manokwari in northwest New Guinea. A similar type of connection has a very much wider use in single outrigger canoes; all of those described from the islands near the eastern end of New Guinea have vertical stanchions supporting the decking from the single outrigger to the dugout.

A fourth type includes booms which are connected to the floats with bent, connecting arms. These arms are ordinarily elbow shaped, the crook of the armpiece being used for attachment to the boom, beyond and below which the long arm of the attachment extends to the float. (See Fig. 36.) There are many variations of this type of connection found in various localities in eastern Java and neighboring islands, such an Menado, Lombok, and Bali, each having its own variety of connecting attachment.

Detailed descriptions merely emphasize the bewildering variety of outrigger canoes. The variety and ingenuity of the attachments from boom to float show how deeply the Indonesians appreciate any improvements which may add to the performance of their double outrigger canoes. The main advantage of these canoes is their great security against upsetting sideways in a sudden gust of wind. Furthermore the space available for cabin and cargo makes them much more convenient than a single outrigger canoe of comparable size.

Many native canoemen of the East Indies seem to the casual foreign observer to be like wild, impulsive children, but there is no mistaking their keen appreciation of a good canoe. They enjoy intensely the tug of the sail to a freshening breeze, the response of the craft to the wind, and the swirl of water left behind.

Perhaps the climax in construction of sailing canoes was reached with the great double canoes which formerly were widely used from New Guinea to Samoa. They were made of two dugout canoes of approximately equal size, up to one hundred feet long and four feet in the beam, with bulwarks raised by a complicated structure of palm midribs lashed together, and supporting a deck from canoe to canoe. They were built at the cost of enormous labor, and displayed great ingenuity. They required additional work to keep them seaworthy and generally were kept in large sheds like airplane hangars.

In Mailu, southern New Guinea, the two dugout hulls are

joined together by cross poles and braced apart by two tenoned poles of equal length, which are mortised into the inner freeboard strakes, one fore and one aft. The cross poles project beyond the main canoe and the outrigger. Even when the canoes are of almost equal size, one is always chosen by the builders to be the canoe proper, and the other is designated as the float. The mast is stepped amidship in the main canoe, and a narrow outboard platform is built upon the projecting poles to give space for the crew to stand when hoisting the sail. A deck or platform covers the space above and between the canoes. No platform is built on the projecting poles on the outrigger side; on them are laid the long oars, poles, and rolled up roofing mats. The mast is made from a strong sapling with the butt ends of its lateral roots (four when obtainable) allowed to remain. By means of these roots the mast is lashed securely into place. The top of the mast is braced to the canoe and to the outrigger by means of twisted cane shrouds. Large rudders, with wide rounded blades and stems about fifteen feet long, are shipped at the stern.

Farther west, along the coast of the New Guinea mainland, the large sailing canoes, lakatoi, are equipped with two masts apiece and two large crab's-claw sails. (See Fig. 37.) Perhaps one of the most picturesque sights in the world is a fleet of these great sailing canoes under way. The effect of their golden brown sails against the tropical foliage of the shore and the dazzling brightness of the shining sea is really startling.

The sails are made by men, but the sail cloth is made by women. They plait narrow strips of matting, about twelve or fifteen inches wide and up to twenty feet long. All such sail cloth is made from a grasslike flowering rush gathered at river banks. The sails are made with a strong rope border to which the edges of the matting are sewed. The well-known crab's-claw shape is carefully planned and is considered very important. The two peaks of the sail are maintained in position by long gaffs made of spliced saplings which extend the full length of each side. The gaffs are laid along the deck when the sail has to be attached, and they project beyond each end of the canoe. The crew has to get out of the canoe

Fig. 37. Lakatoi, double canoe of Papua.

and stand in shallow water to make these attachments before the sail can be hoisted. So large a sail could be carried only by a craft of tremendous stability; this stability is provided by the very wide beam, due to the space between the two canoes. With such a large sail the canoes can go very fast. When the wind is too strong, the canoemen lower the sail and furl it around the lower gaff. For maximum effect the sail is hoisted with the wide end high above the top of the mast, and the point of the sail is made fast to the foot. With so much sail set, the craft would be in great danger if handled badly. In the old days the mainsheet of the great double canoes was given to the king or chief of the tribe as a mark of distinction and as a compliment to his skill as a sailor. In spite of its great stability, a double canoe can easily be wrecked in rough water unless skillfully handled.

The great pride and esteem that the natives held for these canoes is shown by the names they bore. In the Fiji Islands they were frequently named for notable persons lately deceased. In New Guinea the large sailing crafts of today are named for geographical features — a headland, bay, creek, or village. The community to which the canoe belongs takes good care of its canoe and shows great pride in its performance and appearance.

Custom has required certain designs be invariable regardless of utility. If a Papuan is asked why he makes his sail with the pronounced double points, he replies that sails were always made in that shape by his ancestors.

The social positions of the canoe-makers is generally hereditary and monopolistic, and even the kings or chiefs did not presume to build the great double canoes except by commission issued to the master canoe-maker. In many cases the making of the canoe required years of work, even when the building proceeded strictly according to custom. The master canoe-maker in many cases was chosen for his technical skill, his hereditary claims, and his mastery of the necessary magic.

Many Indonesian and Polynesian canoes are decorated by large ovula shells that are hung from pegs or carved points along the decks and at the bow and stern. No one seems to know why these shells are used, but the custom of using them probably has some ancient mythological or superstitious meaning.

Custom is largely responsible for the maintenance of the large double canoes still in use in southeast New Guinea. The main use of the double canoes is for trading and shell hunting. The Mailu islanders and other Papuans wear a shell arm band above the elbow, a custom on which a whole series of other customs depend. For instance, a young man always wears a leaf under his arm band. If any girl takes a fancy to him, she plucks out the leaf, and a rendezvous is then arranged. No arm band, no rendezvous.

The Papuans collect from coral reefs a certain variety of large mollusks from whose shell they make arm bands. The top of the shell has to be chipped carefully away from the natural opening of the shell, then the outside roughnesses and the inside layers are ground on a whetstone. The labor of making shell rings without a fracture is very tedious, and the shells are scarce; consequently, a collection of arm bands constitutes what a western Papuan calls his "wealth." The shells have been in such demand that a good set of two shell arm bands brings the price of $50 to $60 in American money at Port Moresby.

All the local reefs have been combed so carefully that very few shells remain, and the men take their great sailing canoes and go on trips of hundreds of miles to the less frequented reefs on the chance of finding shells. The search for shells is now the main reason why the sailing double canoes are still being built and operated by the Mailus. Although these people have a tendency to live a rather aimless indolent life, their customs spur them on to work for certain objects not necessary to their actual subsistence, but appealing in some way to their ambitions as men. They are steeped in a lore of magic, nearly all of which has a mythological background, in which human and superhuman beings play a part. Their mythology explains and makes requisite for them a way of living to which they feel compelled to conform. Although this leads them to undertake perilous and difficult voyages — like those of the ancient Trojans — it sustains them spiritually, and the romance of adventure lures them on. Into this life and its philosophy the canoe fits as a necessary, indispensable cultural factor.

The study of canoes involves the story of man himself, his undaunted struggles, his romantic adventures, his heroic achievements, and his failures. Canoes reflect the extremes of man's weakness and strength and his infinite resourcefulness. Whenever men have faced the alternatives of starving or of inventing a watercraft to gain access to a distant food supply, they have been able to put together a useful craft. From the Arctic to the Antarctic, the craft they have built has depended upon natural materials available — skins, reeds, bark, bundles of leaves, or hollowed logs. The form of the canoe has depended on the materials of which it was made, and the operation of the canoe has depended on its form and purpose.

Every canoe must be judged according to the setting in

which the man produced it, and the purposes for which he needed it. That is why in looking at canoes the world over we have examined their natural environment as well as their construction, and their design as well as their operation, in order to appreciate fully their value in terms of human culture.

Glossary

Bilge, the bulging part of a ship's side, near the broadest part of a ship's bottom.

Boom, a long pole or spar, used to extend a sail or to connect an outrigger float to a canoe.

Bow, the forepart or prow of a watercraft.

Bulwark, the siding around a ship, usually above the level of the deck; in reference to canoes, the sides of the hull.

Caulk, to stop leaks between the planking of a boat, usually by driving oakum or some other packing material into the spaces.

Centerboard, a movable keel that can be lowered through a watertight slot in the bottom of a sailing boat so as to prevent leeway.

Coaming, a raised rim around an opening, such as the cockpit of a kayak.

Cockpit, an opening through the deck of a covered canoe or boat.

Cutwater, an edge at the bow of a canoe; it may be part of the stem or an attachment to the stem.

False keel, a lengthwise strip attached to the bottom of a canoe, but not forming an essential part of the framework.

False ribs, the ribs which do not reach entirely to the keel or the gunwale.

Fining, the gradual reduction in width of a canoe toward the bow and stern; a drawing of shape toward a point.

Foot, the lower end, as of a mast or sail.

Fore-and-aft, being or moving in the direction of a boat's length.

Freeboard, the vertical distance between the water level and the lowest part of the gunwale.

Gaff, a boom, pole, or yard to extend the upper edge of a fore-and-aft sail.

Gunwale, the lengthwise strip covering the upper ends of the ribs and the top of the bulwarks in a canoe.

Halyard, a rope for hoisting a sail.

Head, the upper part or top, as of a sail or mast.

Keel, the chief and lowest lengthwise strip in the framework of a canoe.

Lee, the side opposite to that from which the wind blows; the side sheltered from the wind.

Leech, either edge of a square sail; the after edge of a fore-and-aft sail.

Leeway, the drift of a vessel to leeward due to the wind.

Luff, the forward or weather leech of a fore-and-aft sail.

Mast, the pole or group of poles set up in a boat to support the sails.

Port, the left side of a vessel as one looks from the stern toward the bow.

Portage, the carrying of canoes and goods overland from one watercourse to another.

Rake, an inclined position.

Ribs, the upright side frames of a canoe between the keel and the gunwale, holding in place the side sheathing and lengthwise strips.

Rudder, a broad paddle-like device usually hinged at the stern of a boat in order to guide the course; in canoes large paddles are commonly used in place of rudders.

Sheathing, the protective lining or covering of a canoe's hull.

Sheer, the upward curve of the lines of a vessel's hull, keel, or gunwales.

Sheet, a rope attached to the lower corner of a sail by which the sail is controlled.

Ship, to put in place on a canoe, such as on a mast or a rudder.

Starboard, the right side of a vessel looking from stern to bow.

Stem, a nearly upright piece in the framework of a canoe which forms the bow.

Step, to insert the lower end of a mast in a socket or other suitable fastening.

Stern, the aft part of a canoe, opposite to the fore or front end.

Stern sheets, a seat or platform in the stern above the canoe bottom and below gunwale level.

Strakes, a breadth of planking on a ship's hull.

Tack, to change the course of a sailing vessel so as to bring the wind on the other side, by turning the bow of the ship through an arc in such a way that it points momentarily straight into the wind in making the turn.

Tholepin, a wooden pin set in the gunwale of a boat to serve as a fulcrum for the oar.

Thwart, a stake or seat stretching across a canoe from gunwale to gunwale or from side to side.

Tiller, a lever for turning the rudder.

Tumble home, a narrowing of the width of a canoe toward the gunwales.

Wash strake, a strake raised above the gunwale level to keep out wave splash.

Wave-form hull, one which has a marked sheer of the lower lines toward both bow and stern.

Wear, to change the course of a sailing ship in such a way that it points momentarily directly away from the wind in making the turn.

Windward, the direction from which the wind is blowing.

Yard, a pole or spar set crosswise to a mast for the support of a sail.

Bibliography

Adney, Edwin Tappen. "The Building of a Birch Canoe," *Outing*, XXXVI (1900), 185-89.

Alexander, A. B. "Notes on the Boats, Apparatus and Fishing Methods Employed by the Natives of the South Seas," *House Doc.*, Vol. 127 (1901-02), 57th Congress of the United States.

Balfour, Surgeon General Edward. *Cyclopedia of India*, 3d ed., London: 1885, pp. 392-400.

Barrett, O. W. "Impressions and Scenes of Mozambique," *National Geographic*, XXI (1910), 809-17.

Barrett, S. A. "Cayapá Indians of Ecuador," *Indian Notes and Monographs*, Vol. 40 (1925), Pt. 1.

Brandes, E. W. "Into Primeval Papua by Seaplane," *National Geographic*, LV (1929), 253-332.

Buck, Peter Henry. *The Material Culture of the Cook Islands*. New Plymouth, N. Z.: 1927.

Burke, Walter. "Hurdle Racing in Canoes," *National Geographic*, XXXVII (1920), 440-44.

Buschan, George. *Illustrierte Völkerkunde Stuttgart*, I (1922), 260-61.

Camsell, Charles. "Some Interesting Geographical Problems in the Exploration of Northern Canada," *Geographical Review*, V (1918), 208-16.

Cheesman, Major R. E. "The Highlands of Abyssinia," *Scottish Geographic*, 52 (1936), 8.

Chevrillon, Andre. "Cavelier de La Salle jusqu'à la prise de possession de la Louisiane, 1643-1682," *Rice Inst. Pamph.*, XXIV (1937), 119-42.

Church, John W. "A Vanishing People of the South Seas," *National Geographic*, XXXVI (1919), 275-306.

Churchill, W. M. "Samoan Canoes," *Outing*, XXXV (1900), 75-77.

Clemens, W. M. "With Sail and Paddle in the South Seas," *Outing*, XXXVII (1901), 568-72.

Farabee, W. C. "The Central Arawaks," *The Univ. Mus. Anthropol. Publ.*, University of Pennsylvania, Vol. IX (1918).

Flandreau, Grace. "Then I Saw the Congo," *National Geographic*, LXII (1937), 657.

Fosbrooke, H. A. "Some Aspects of the Kimwani Fishing Culture," *J. Royal Anthropol. Inst. Gr. Brit.*, LXIV, New Ser. 37 (1934), 10-11.

Haddon, A. C. "The Outrigger Canoe of East Africa," *Man*, No. 29, April (1918).

———. "An Anomalous Form of Outrigger Attachment in Torres Straits and Its Distribution," *Man*, No. 68, Aug. (1918).

Hambly, Wilfred D. *Field Mus. Nat. Hist. Anthropol. Ser.*, 21 (1931), Plates CXLIII and LXXXV, 145-46.

Hawkes, E. W. "The Labrador Eskimo," *Geol. Surv. Canada Mem. No. 91* (1916).

Heizer, Robert F. "The Plank Canoe of the Santa Barbara Region, California," *Ethnologiska Studier,* VII (1938), 193-227.

Hornell, James. "African Bark Canoes," *Man,* No. 198 (1935).

———. "The Boats of the Ganges," *Mem. of the Asiatic Soc. Bengal,* Vol. VIII (1924), No. 3.

———. "The Outrigger Canoes of Indonesia," *Madras Fisheries Dept. Rpt. No. 2,* XII (1920), 43-114.

———. "Outrigger Canoe of Madagascar," *Asia,* XXX (1930), 168-70.

Howley, James P. *The Beothucks or Red Indians.* Cambridge: Cambridge University Press, 1915.

Jenness, D. *The People of the Twilight.* New York: 1928.

Johnston, Sir Harry. *George Grenfell and the Congo.* London: 1908.

———. *The Uganda Protectorate.* London: 1902.

Kahn, Morton C. *Djuka.* New York: 1931.

Kindle, E. M. "Notes on the Forests of Southeastern Labrador," *Geographical Review,* XII (1922), 57-71.

Lloyd, T. G. B. "The Beothucks," *J. Royal Anthropol. Inst. Gr. Brit.,* IV (1874), 26-28.

Lothrop, S. W. "Indians of Tierra del Fuego," *Mus. Amer. Ind. Heye Found.,* X (1928), 143-45.

———. "Aboriginal Navigation off the West Coast of South America," *J. Royal Anthropol. Inst. Gr. Brit.,* LXII, New Ser. 35 (1932), 229-56.

MacRitchie, David. "Kayaks of the North Sea," *Scottish Geographic,* XXVIII (1912), 126-33.

———. "The Eskimos of Davis Straits in 1656," *Scottish Geographic,* XXVIII (1912), 281-94.

Malinowski, Bronislaw. *Argonauts of the Western Pacific.* London and New York: 1922.

Marquez, Luis, and Helene Fischer. "Photographs of Tarascan Canoemen," *National Geographic,* LXXI (1937), 637-44.

Mason, Otis T. "Pointed Bark Canoes of the Kootenai and Amur," *Rpt. of U. S. Nat. Mus., 1899.* Washington: 1901, pp. 525-37, Plates I to V.

———. "Aboriginal American Canoes," *Outing,* XLIV (1904), 704-13.

Mead, Margaret. *Growing up in New Guinea.* New York: 1930.

Mitman, Carl W. "Catalogue of the Watercraft Collection in the U. S. National Museum," *U. S. Nat. Mus. Bul. 127* (1923).

Moore, W. Robert. "Raft Life on the Hwang Ho," *National Geographic,* LXI (1932), 743-52.

Pitt-Rivers, A. Lane-Fox. *The Evolution of Culture.* Oxford: 1906.

Ravenstein, E. G. *The Russians on the Amur.* London: 1861.
Roop, Wendell P. *Watercraft in Amazonia.* Woodbury, N. J.: 1935.
Roth, A. E. "Indians of Guiana," *Bur. Amer. Ethnol. 38th Ann. Rpt.* (1924).
Savelle, W. J. V. *In Unknown New Guinea.* London: 1926.
Skinner, Alanson. "Material Culture of the Menominee," *Indian Notes and Monographs,* Vol. 20 (1921).
Skinner, H. D. "Moriori Seagoing Craft," *Man,* No. 34 (1919).
Smith, Elliot G. "Ancient Mariners," *J. Man. Geog. Soc.,* XXXIII (1917), 1-20.
Steinen, Karl von den. *Durch Zentral-Brasilien.* Leipzig: 1886.
———. *Unter den Naturvölkern Zentral-Brasiliens.* Berlin: 1897.
Stiere, J. B. "Indian Tribes of Puru River," *Rpt. of U. S. Nat. Mus.,* 1901, p. 359.
Stirling, Matthew W. "America's First Settlers, the Indians," *National Geographic,* Vol. LXXII (1937).
Suder, Hans. "Vom Einbaum und Floss zum Schiff," *Institute für Meereskunde.* Historical-Anthropological Series, Vol. 7 (1930).
Thomas, N. W. "Australian Canoes and Rafts," *J. Royal Anthropol. Inst. Gr. Brit.,* XXXV (1905), 56-79.
Thurn, E. F. im. *Among the Indians of Guiana.* London: 1883.
Torday, Emil. *Camp and Tramp in African Wilds.* London: 1913.
———. "Curious and Characteristic Customs of Central African Tribes," *National Geographic,* XXXVI (1919), 342-68.
Waterman, T. T., and Geraldine Coffin. "Types of Canoes on Puget Sound," *Indian Notes and Monographs,* Vol. 1-6 (1920).
Waugh, F. W. "Canadian Aboriginal Canoes," *The Canadian Field Naturalist,* XXXIII (1919), 23-33.
Whitney, Caspar. "Through the Klungs of Siam," *Outing,* XLIV (1904), 420-35.
Willoughby, C. C. *Antiquities of the New England Indians.* Peabody Museum, Harvard University, 1935.
Willson, Beckles. *The Great Company.* Toronto: 1899.
Zenzinov, Vladimir M. "With an Exile in Arctic Siberia," *National Geographic,* XLVI (1934), 695-718.

Index

Aberdeen, 68
Abyssinia, 57n
Acadia, 45
Africa, 21, 24, 25, 26, 28, 33, 35, 36, 37, 52, 53, 55, 57, 58, 86, 117
Africans, 23, 26, 27, 33, 40, 58; African canoemen, 21, 29, 33, 37, 40; canoes, 37, 87; dugout, 36; paddle blade, 29; paddlers, 27, 28; slaver, 41; tribes, 38, 40, 53; villages, 39; yarn, 41
Alaska, 11, 24, 33, 43, 47, 66, 69, 77, 99, 101; expedition to, in 1919, 69
Alaskan, kayak, 68; models, 68; umiak, 71; waters, 71
Aleutians, 23, 31, 70, 71, 106; Aleutian canoemen, 69; Eskimo, 107; skin canoes, 106
Aleutian Islands, 66, 69, 70, 71
Algonquians, 90
Amazonians, 51; Amazonian canoemen, 51; canoes, 10; Indians, 20, 51, 85; tribes, 51; "uba," 50, 51
Amazon River, 28, 50, 84, 85
Amboina, island of, 119
Americans, 33, 70, 119; American birchbark canoe, 105; elm, 91; models, 105; pilot boat, 108
Amur, canoemen, 23; canoes, 106; tribes, 106
Amur River, 91, 104, 105, 106
Andes Mountains, 60, 62
Angola, 37, 86
Annam, 74
Antarctic Ocean, 79, 124
Anula tribe, 83
Arabs, 33
Arawak Indian, 17; word, 9

Arctic, 10, 23, 29, 33, 68, 72, 124; animals, 66; Eskimos, 107; kayak, 33; seas, 66, 69
Arctic Circle, 98
Arctic Ocean, 33, 106
Arctic Russia, 71
Arctic Siberia, 17, 23
Argentina, 51, 52, 74
Arkansas River, 47
Asia, 33, 77, 107
Asiatics, 23, 33; Asiatic canoemen, 104; canoes, 107; tribes, 104
Asiatic Russia, 104
Assyria, 77
Athapascan, canoe, 99; Indians, 99, 100
Atlantic Coast, 52
Australasia, 63
Australia, 14, 15, 33, 62, 82, 83, 117
Australians, 15, 62, 82, 83; Australian sea hunter, 81

Bakwese, 58
Bali, 117, 120
Bengal, 16, 74
Beothucks, 99, 101, 102, 103; Beothuck canoe, 103, 107
Bering Sea, 31, 33
Biraris, 104, 105; Birari canoes, 105, 107
Blue Nile, 57, 74; *see also* Nile River
Bolivia, 31, 50, 60, 84
Bombay fishing canoe, 109
Borneo, 118
Bradbury, C. Earl, 7, 8
Brazil, 13, 31, 33, 50, 81, 84, 85, 86
Brazilian woodskin, 87, 95

131

Britain, 20
British, 33
British Columbia, 27, 44, 47
British Guiana, 84, 86
Britons, 75
Buccaneer Archipelago, 15
Buduma tribe, 57
Burma, 20, 29
Burroughs, Stephen, 66
Bush Negroes, 12

Caesar, Julius, 20, 75
California, 44
Canada, 7, 92, 94, 101, 106; Geological Survey of, 7
Canadian, canoe, 28; lumberjack, 14
Cangamba, 86, 87; woodskin, 87
Carib Indians, 17
Cayapa, dugout, 49; Indians, 17, 49, 50, 53
Central America, 47, 48
Ceram, island of, 119
Ceylon, island of, 111; craft, 112; single outrigger, 112
Chatham Islands, 63
Cheesman, Major R. E., 57
Chesapeake Bay, 45
Chicago, 94
Chile, 24, 78, 79
Chilean paddler, 11
China, 25, 29, 74, 77
Chinese, dragon-boat races, 11; records, 20
Chipewyan, 100
Christian, Era, 117; monasteries, 57
Clark; *see* Lewis and Clark
Coeur d'Alene Indians, 92
Colombia, 49, 50
Columbia River, 92, 93
Columbus, 9

Congo, 12, 22, 27, 35, 36, 37, 40, 41, 53, 58; tribes, 24
Congo River, 36, 57
Connecticut, 45
Cook, Captain James, 44, 66, 70
"Corwen," 71
Crees, 101
Creole frog hunters, 22, 46

Danish, expedition, 1653, 71; settlements, 72
Dibil palm, 82
Dinkas, 34, 38; Dinka canoemen, 38; canoes, 38
Djukas, 12, 52, 53; Djuka canoe, 52, 53; canoemen, 52, 53; paddle, 53
Dog-ribs; *see* Athapascan Indians
Dutch East Indies, 25, 118
Dutch government, 118, 119
Dutch Guiana, 52

Easter Island, 11
East Indies, 120
Ecuador, 17, 25, 49, 50, 52, 53; Indians, 50
Egypt, 54, 56
Egyptians, 9, 12, 20, 55, 58; Egyptian craft, 55; papyrus floats, 13
England, 7
Equator, 24
Eric the Red, 66
Eskimos, 10, 11, 20, 23, 24, 29, 33, 65, 66, 67, 68, 70, 72, 77, 82, 98, 100; Eskimo canoeman, 12, 23, 72; Christianized, 72; deck covering, 99; kayak, 77, 99, 105
Ethiopia, 54, 57
Ethiopians, 74
Euphrates, 74

Europe, 13, 22, 31, 75, 108
Europeans, 24, 27, 29, 33, 52, 62, 104; European bargee, 22; boatbuilder, 18; boats, 109; explorers, 35; frigates, 108; gunboats, 108; sailing ships, 36; sailors, 111; warships, 20
Everglades, 47

Far East, 104
Fernando Poo, 33
Fiji Islands, 123
Flood, the, 72
Florida, 22, 47
Fort Chipewyan, 94
French fur traders, 101
Funk Island, 103

Gilbert Islands, 11, 116
Giliaks, 104
Goldis, 104, 105; Goldi canoes, 106
Grabham, George, 38n
Gran Chaco, 51, 74
Greater Pacific, 24
Great Lakes, 101
Great Slave Lake, 100
Great Spirit, 13
Greeks, 9, 20
Green Bay, 90
Greenland, 66, 68, 72; Eskimos, 67, 69; kayak, 65, 68
Greenlanders, 69
Guatemala, 48
Guayaguil jangada, 32
Guiana, 12, 50; cedar, 53; Indians of, 17
Guinea Coast, 36, 58
Gulf of California, 59
Gulf of Carpentaria, 83
Gulf of Mexico, 94

Gulf Stream, 72
Gunnbjörn, 66

Haida Indians, 18; dugout canoe, 42
Hawaii, 9
Hawaiian islander, 11
Hay River, 100
Herodotus, 74
Himalayas, 74, 76
Homer, 9
Hudson Bay Post, 96, 97
Hudson's Bay Company, 97, 101
Huron, country, 100; Indians, 101; Lake, 101

Incas, 13
India, 16, 74, 77; fishermen of, 58
Indian Ocean, 87, 112
Indians, 12, 13, 24, 31, 33, 42, 44, 45, 49, 52, 53, 73, 74, 77, 78, 84, 86, 91, 92, 93, 94, 96, 97, 98, 102, 104, 110; Indian canoemen, 31; canoes, 10, 52, 99; paddles, 53; plains, 73; seagoing, 111; woodland, 29, 33, 43, 73, 90, 94, 101
Indonesians, 118, 120, 123; Indonesian canoes, 33
Indus River, 58, 59
Iraq, 74, 77
Ireland, 75
Irish, curragh, 75; peasantry, 75
Iroquois, canoe, 90; country, 100; Indians, 90, 101
Isaiah, 54
Isis, 55

Japan, 31, 104, 115
Jappen Islands, 116

Java, 116, 118, 119, 120
Javanese, 117
Johore fast boat, 11, 108, 109

Kasai River, 58
"Kellek," 77
Khartoum, 16
Kimwani tribe, 25, 26
Kiwanga, Towegale, 40
Kootenai, canoe, 27, 91, 92, 93, 104; Lake, 93; tribe, 93
Kru, 36, 37
"Kula Ring," 114, 115
Kutchin tribe, 77
Kwando River, 86

Labrador, 69, 98, 101; Eskimo, 69; kayak, 69
Ladrones, 115
Lake Chad, 22, 57
Lake Moeris, 56
Lake Patzcuaro, 48
Lake Tana, 56, 57
Lake Tanganyika, 33
Lake Titicaca, 31, 33, 59, 60, 62, 111
"Land of Fire"; see Tierra del Fuego
Lapland, 33
Laplanders, 68
La Salle, Cavelier de, 94
Lewis and Clark, 44
Lloyd, T. G. B., 77, 78, 102, 103, 104
Lol River, 34
Lomami, 58
Lomami River, 37
Lombok, 120
Lothrop, S. W., 80
Louisiana, 22, 46
Lualaba River, 41

Mackenzie, 94
Mackenzie River, 69, 77, 99
Madagascar, 11, 33, 37, 40
Madura, 119
Magellan, 115
Mailu, island of, 120; islanders, 123, 124
Maine, 101
Makassar Strait, 116
Malaya, 25, 29, 108
Malayan pirates, 108
Malecite Indians, 96, 100
Manchuria, 91
Manokwari, 119
Manyorgs, 104, 105; Manyorg canoes, 106
Maoris, 54, 63; Maori war-canoe, 19
Mariana Islands, 115
Marquesas, 118
Marquette, Père, 94
Marshall Islands, 30, 116
Mediterranean, 9
Menado, island of, 119, 120
Menominee, 90, 91, 97
Mesopotamia; see Iraq
Mexican Indians, 48
Mexico, 26, 44, 47, 59
Michilimackinac, 94
Micmac Indians, 77, 98, 101
Micronesia, islands of, 115, 116
Middle West, 46, 98
Minnesota, 29
Mississippi River, 94
Mongolia, 94, 104, 106
Mongolian canoeman, 107
"Monitor," 91
Montagnais, 100
Moriori, 63, 64
Moses, 54, 55

Mozambique, 87, 88; canoe, 87, 88
Museum of Anthropology, 88

Nansen, Captain Fridtjof, 72
National Geographic magazine, 8
National Geographic Society, 69
Negro, slaves, 52; tribes, 35
Nevada, 59
New Brunswick, 98, 100
New England, 29
Newfoundland, 77, 101, 102, 104
New Guinea, 28, 111, 112, 114, 116, 117, 118, 119, 120, 121, 123
New York, 90
New Zealand, 11, 19, 29, 55, 63
Ngombe hippopotamus hunter, 40
Nigeria, 37
Nigerian ceremonial dances, 24
Niger River, 36, 57
Nightingale, E. H., 38n
Nile River, 16, 36, 37, 54, 57
Nipigon region, 95
"Noah's wood," 72
Norsemen, 66
North America, 10, 18, 24, 42, 44, 47, 62, 91, 107
North American, bull-boat, 74; canoemen, 25, 29; hold, 27, 50; Indian, 10, 11, 50, 51, 59, 77, 107; paddle blade, 28; tribes, 25
North Carolina, 45
Nyangas, 33

Oceania, 63
Odyssey, 9
Ojibways, 95, 101, 106; Ojibway canoes, 107
Ontario, 7, 100; Lake, 101

Orochi, 104
Osiris, 55

Pacific Coast, 42, 52
Pacific Ocean, 11, 31, 94
Papuans, 21, 28, 123, 124; Papuan double canoes, 122; paddlers, 28; paddles, 21
Paraguay, 74, 84
Parana canoemen, 51
Parana River, 51
Paru River, 86
Passamaquoddy tribe, 101
Pequot Indians, 45
Persian Gulf, 74
Peru, 31, 33, 50, 52, 110
Peruvian, coast, 24; harbors, 60; Indians, 52; rafts, 110
Petchora River, 68
Philippine Islands, 117, 119
Pliny, 55
Plutarch, 55
Poland, 22
Polynesia, 9, 25, 27, 33
Polynesians, 23, 29, 113, 114; Polynesian canoes, 123; clothing, 114; paddles, 29; sailors, 11
Port Moresby, 124
Portuguese East Africa, 87
Praeneste, 55
Pripet marshes, 22
Pungwe River, 87
Punjab, 77
Pygmies, 35
Pyramid Lake, 58, 59, 60

Quebec, 100
Queensland, 82, 83
Quirke, Terence T., 7
Quirke, Terence T., Jr., 8

Red Indians, 102
Red River, 69
Red Sea, 28, 37
Restigouche River, 98
Rhodesia, 87
River Boyne, 75
Romans, 9, 20
Rome, 55
Rongotakuiti, 64
Russia, 66, 68
Russian traders, 31, 33, 71

Sakhalin Island, 104
Salish tribe, 93
Samoa, 120
Samoyeds, 17, 23, 68
Saulteaux, 95, 100; Saulteaux wigwams, 95
Scandinavians, 33
Scotch peasantry, 75
Scotland, 75
Scottish Geographic, 57
Scottish waters, 68
Seminole Indians, 22, 47; dugout, 46
Senegambia, 58
Shilluk, canoemen, 38; papyrus canoe, 54, 55; tribe, 16, 38, 39, 55, 56
Siamese dragon-boat races, 11
Siberia, 91, 94, 104, 106, 107
Siberian, canoe, 107; canoemen, 107; tribes, 106
Singapore, 31
Slave Indians, 100
Smithsonian Institution, 8
Solomon Islands, 118
Somali, 28
South America, 17, 24, 27, 31, 32, 33, 49, 59, 62, 74, 78, 80, 96, 110

South American, aborigines, 111; bark, 85; kayak, 78; raft, 11; rivers, 52
South Seas, 33, 118; South Sea canoemen, 114; islander, 116
Spaniards, 9, 31
von den Steinen, Karl, 84
Stevens, Neil E., 45
Straits of Magellan, 88, 89
Sudan, 16, 34, 38
Sumatra, 118
Sutlej River, 76
Syrian, 77

Tahltan Indians, 77
Tarascan Indians, 26, 28, 48; dugouts, 47; fishermen, 47
Tasmania, 62
Tasmanians, 62
im Thurn, E. F., 84, 85
Thuron Island, 59
Tibet, 74
Tibetan stream, 73
Tierra del Fuego, 10, 88, 89, 90
Tigris, 74
Tinneh Indians; *see* Athapascan Indians
Trobriand islanders, 112, 116
Trojans, 124
Tungus, 104, 105, 106

Ubangi River, 22, 25
Ubena tribe, 21, 36
Ulindi River, 40
Ulysses, 9
United States, 33, 45, 49, 92; Fish and Wildlife Service, 8; National Museum, 43, 71, 75, 76
Uruguay, 52

Vachokwe tribe, 86
Vaigach, 66
Venezuela, 50, 84, 86
Victoria Nyanga, 25
Virginia, 45

Waito sailors, 57
Watet, 23, 25
West Indies, 9
Wisconsin, 90

Xingu River, 84, 86

Yahgans, 10, 88, 89, 90, 95;
    Yahgan canoes, 95
Yakut, canoemen, 106; tribe, 106, 107
Yangtze Kiang, 74
Yankee clipper ship, 108
Yapongu tribe, 37
Yukon River, 99
Yukon Territory, 77

Zambezi basin, 87